WHEN SOLDIERS SAY NO

Military and Defence Ethics

Series Editors

There is an urgent and growing need for all those involved in matters of national defence – from policy makers to armaments manufacturers to members of the armed forces – to behave, and to be seen to behave, ethically. The ethical dimensions of making decisions and taking action in the defence arena are the subject of intense and ongoing media interest and public scrutiny. It is vital that all those involved be given the benefit of the finest possible advice and support. Such advice is best sought from those who have great practical experience or theoretical wisdom (or both) in their particular field and publication of their work in this series will ensure that it is readily accessible to all who need it.

Also in the series

From Northern Ireland to Afghanistan
British Military Intelligence Operations, Ethics and Human Rights
Jon Moran
ISBN 978-1-4094-2897-8

Military Medical Ethics for the 21st Century
Edited by Michael L. Gross and Don Carrick
ISBN 978-1-4094-3898-4

Precision-guided Munitions and Human Suffering in War
James E. Hickey
ISBN 978-1-4094-2951-7

When Soldiers Say No

Selective Conscientious Objection in the Modern Military

Edited by

ANDREA ELLNER
King's College London, UK

PAUL ROBINSON
University of Ottawa, Canada

DAVID WHETHAM
King's College London, UK

Routledge
Taylor & Francis Group
LONDON AND NEW YORK

First published 2014 by Ashgate Publishing

2 Park Square, Milton Park, Abingdon, Oxon OX14 4RN
711 Third Avenue, New York, NY 10017, USA

Routledge is an imprint of the Taylor & Francis Group, an informa business

First issued in paperback 2016

British Library Cataloguing in Publication Data
A catalogue record for this book is available from the British Library

The Library of Congress has cataloged the printed edition as follows:
Ellner, Andrea.
When soldiers say no : selective conscientious objection in the modern military / by Andrea Ellner, Paul Robinson, and David Whetham.
pages cm. – (Military and defence ethics)
Includes bibliographical references and index.
ISBN 978-1-4724-1214-0 (hardback)
1. Selective conscientious objection. I. Robinson, Paul, 1966– II. Whetham, David. III. Title. IV. Title: Selective conscientious objection in the modern military.

U22.E465 2013
174'.9355–dc23

2013021068

ISBN 978-1-4724-1214-0 (hbk)
ISBN 978-1-138-24605-8 (pbk)

Contents

PART II CASE STUDIES IN SELECTIVE CONSCIENTIOUS OBJECTION

PART III CONCLUSIONS

Notes on Contributors

Richard Adams is a serving officer in the Royal Australian Navy. He was educated at the University of Tasmania (BA), the University of New South Wales (MA) and the University of Western Australia (B.Ed. Hons, MIR and PhD). He was a Fulbright Scholar to Yale University, where he was a Visiting Fellow on the Global Justice Program. Posted presently to the Centre for Defence Leadership and Ethics within the Australian Defence College, Richard is a Visiting Fellow to the School of Humanities and Social Science at the Australian Defence Force Academy, University College, the University of New South Wales.

Melissa Bergeron is an assistant professor of philosophy at the United States Military Academy, West Point, where she teaches philosophy, composition, and professional military ethics. Her areas of specialization include epistemology (especially, the ethics of belief) and moral theory, in particular just war theory.

Nikki Coleman is a postgraduate research student, examining ethical problems related to the conditions of military service, in the School of Humanities and Social Sciences, University of NSW, Canberra, at the Australian Defence Force Academy. She has taught in the School's military ethics program for a number of years and is also an ordained minister in the Uniting Church in Australia.

Stephen Coleman is senior lecturer in Ethics and Leadership in the School of Humanities and Social Sciences, and Program Director, Military Ethics at the Australian Centre for the Study of Armed Conflict and Society with the University of NSW, Canberra, at the Australian Defence Force Academy. He has published on a diverse range of topics in applied ethics, including military ethics, police ethics, medical ethics, and the practical applications of human rights. His latest book, *Military Ethics: An Introduction with Case Studies* was published by Oxford University Press in 2012.

Stephen Deakin teaches at the Royal Military Academy Sandhurst. He also teaches leadership and management to senior military leaders around the world as part of the British defence diplomacy programme. Stephen researches in the areas of military and professional ethics and has published widely in these areas.

Andrea Ellner joined the Defence Studies Department of King's College London in 2007. Prior to this she lectured for nearly ten years on International Security Studies and related subjects at the University of Reading, where she also led the

Graduate Institute of Political and International Studies for three years. Dr Ellner received her MA in History from the Ludwig-Maximilians-University, Munich, and her PhD in Political Science on British Naval Policy, 1970-1990, from the Free University of Berlin.

J. Carl Ficarrotta recently retired after many years as a professor of philosophy at the United States Air Force Academy and is currently a visiting scholar at the University of North Carolina at Chapel Hill. He is the author of *Kantian Thinking about Military Ethics* (Ashgate, 2010).

David Fisher is a teaching fellow at King's College, London in the Department of War Studies. Before this, he was a senior official in the Ministry of Defence, Foreign Office and Cabinet Office, serving as defence adviser to the Prime Minister and UK Defence Counsellor to NATO. He is co-Chairman of the Council on Christian Approaches to Defence and Disarmament and regularly contributes to books and journals on defence and ethics. His most recent book is *Morality and War: Can War be Just in the Twenty-first Century?* (Oxford University Press, 2011) which won the 2012 W.J. Mackenzie prize for the best book of the year in political science.

Emmanuel Goffi serves in the French Air Force. He lectured in International Relations and the Law of Armed Conflict at the French Air Force Academy for five years before he was appointed as a research associate at the Center for Aerospace Strategic Studies in Paris. He is currently doing a PhD (ABD) in Political Science at the Institute of Political Studies in Paris (SciencesPo-CERI). Captain Goffi also teaches international relations at the DSI Diplomatic Studies Institute in Marseille. He is the author of *Les armées françaises face à la morale: une réflexion au cœur des conflits modernes* (L'Harmattan, 2011).

Lieutenant Colonel Brian Imiola is academy professor of philosophy in the Department of English and Philosophy at the US Military Academy West Point. LTC Imiola has served as both an Infantry and Logistics officer prior to his assignment as a permanent member of West Point's senior faculty. He has deployed to Somalia, Haiti, and Iraq. His recent scholarship has focused on exploring the Professional Military Ethic.

Yves LeBouthillier is a professor of law at the University of Ottawa, where he teaches courses on international environmental law, immigration and refugee law and public law. He is currently director of the IUCN Academy of Environmental Law.

Jeff McMahan is professor of philosophy at Rutgers University. He is the author of *The Ethics of Killing: Problems at the Margins of Life* (2002) and *Killing in War* (2009), both published by Oxford University Press. A collection of his essays in bioethics, *The Values of Lives*, is forthcoming from Oxford University Press.

Yossi Nehushtan teaches at the Haim Striks School of Law in Israel and at Keele University, School of Law. He received a doctorate in Philosophy at the University of Oxford and his main research and teaching fields are Public Law, Human Rights Law, Legal and Political Theory and Law and Religion.

Paul Robinson is a professor in the Graduate School of Public and International Affairs at the University of Ottawa. He is the author and editor of numerous works on military history and military ethics, including *Military Honour and the Conduct of War: From Ancient Greece to Iraq* (Routledge, 2006).

Jürgen Rose is a recently retired Lieutenant Colonel of the Bundeswehr and German publicist. He gained a public profile in 2007, when he successfully requested to be relieved from military duties relating to the conflict in Afghanistan on grounds of conscience. He is the author of *Ernstfall Angriffskrieg. Frieden schaffen mit aller Gewalt?* (Ossietzky Verlag, 2009) and has published extensively on a wide range of military, defence, and security policy issues.

Michael Skerker is assistant professor in the Department of Leadership, Ethics and Law at the US Naval Academy. He previously taught at the University of Chicago and at DePaul University. His academic interests include professional ethics, just war theory, moral pluralism, theological ethics, and religion and politics. Publications include works on ethics and asymmetrical war, moral pluralism, intelligence ethics, and the book *An Ethics of Interrogation* (University of Chicago Press, 2010).

David Whetham is senior lecturer in Defence Studies at King's College London, based at the Joint Services Command and Staff College at the UK Defence Academy. Publications include *Ethics, Law and Military Operations* (Palgrave, 2010) and *Just Wars and Moral Victories: surprise, deception and the normative framework of European war in the later Middle Ages* (Brill, 2009). David sits on the Editorial Advisory Board of the Journal of Military Ethics, and is Vice President of the European Chapter of the International Society of Military Ethics (Euro ISME).

Daniel Zupan graduated from the University of Montana with a B.A. double-major in Philosophy and Spanish. After graduation, he joined the US Army, attended Officer Candidate School, and became an infantry officer. He served for 27 years with assignments in Europe, the Continental United States, and Afghanistan. Most of his operational time was with the 82d Airborne Division. He earned an MA in Philosophy from the University of Wisconsin-Madison and a PhD in Philosophy from the University of New Mexico-Albuquerque and went on to teach philosophy at the United States Military Academy (USMA) at West Point, NY. At the time of his retirement as a Colonel, he was head of the philosophy program at USMA. Dr Zupan currently works for the US Army's Human Terrain System where he is the Chief of Leadership Instruction.

Foreword

Jeff McMahan

According to international law and international criminal law, soldiers do not act illegally by fighting in an illegal war, provided they obey the laws that govern the conduct of war (*jus in bello*). Similarly, and presumably on the basis of the same considerations that have shaped the law of armed conflict, the traditional theory of the just war holds that soldiers do no wrong by fighting in a war that is unjust, provided they do not violate the moral principles of *jus in bello*. If the traditional theory is right, most conscientious objectors are simply mistaken. For if fighting by the rules is always morally permissible, even in a war that is unjust, then there can be no general moral basis for refusing to fight, and hence nothing for a person's conscience to object to. While it is obvious that just war theory rejects the pacifist claim that morality requires conscientious refusal to participate in any war, it is surprising that it also denies that there is any general moral objection to participation in wars that fail to meet the conditions of a just war and thus, by implication, denies that there is any general moral basis for selective conscientious objection to participation in wars that are impermissible by the theory's own lights. But, while this seems a clear implication of the traditional theory, I know of no traditional just war theorists who have called attention to it. (When I say that traditional just war theory implies that there is no *general* moral basis for selective conscientious objection, I mean to allow for the possibility that a person may have a moral reason for refusing to fight that is specific to his individual situation, such as that he has promised his mother that he will never fight in an unjust war. We need not consider such possibilities here.)

In recent years many philosophers, myself included, have argued that traditional just war theory is itself mistaken. They have argued that the moral permissibility of fighting in a war cannot be divorced from the aims of the war. Consider, for example, a soldier who is fighting in a war that is intended to achieve only aims that are unjust. Suppose that he is not under threat, so that his purpose in firing his weapons is not self-defense but only to contribute to winning the war. He is therefore killing soldiers on the opposing side, who have done nothing wrong, as a means of achieving unjust aims. His reason for killing them is only that they are attempting to defend themselves and others from unjust aggression – that is, he is killing them as a means of preventing them from justifiably obstructing the achievement of unjust aims. It is difficult to see how that could be morally permissible.

Just war theorists often say that the reason it is permissible to fight in an unjust war is that the moral principles that govern the resort to war, the principles of *jus ad bellum*, apply only to states and their authorized leaders, not to individual soldiers. That is, it is only the state that may not fight if the war would be unjust. That it is impermissible for the state to fight is supposed to be compatible with its being permissible for the soldiers to fight (though what the difference is between their fighting at the command of the state and the state's fighting is rather obscure). The parallel claim about the law of *jus ad bellum* – that it is illegal for the state to fight but legal for its soldiers to fight – is certainly true, for that law was intentionally designed to give soldiers immunity from legal liability to punishment as long as they fight by legally permissible means, no matter how obviously illegal their war may be. In the world as it is, there are sound reasons to grant this conditional immunity. For immunity from punishment, which combatants forfeit if they violate the laws of *jus in bello*, is intended as an incentive for adherence to those laws. The law in effect offers combatants a deal: even if the war in which they are fighting is criminal, it offers them immunity from punishment, as well as humane treatment if they are captured, in exchange for their obedience to certain rules, the most important of which is the prohibition of intentional attacks on civilians. But this and other related reasons for granting combatants legal immunity from punishment for fighting in an unjust and criminal war are not reasons for supposing that fighting in such a war is morally permissible. While morality may dictate that in certain circumstances we design and enforce certain rules with the aim of manipulating people's behavior through the use of incentives and penalties, it is not itself a set of such rules.

Another often proposed explanation of why it is permissible to fight in an unjust war is that responsibility for the war, and for the soldiers' participation in the war, lies exclusively with the state's political leaders. A soldier is, in Augustine's words, "but the sword in the hand of him who uses it, [and so] is not himself responsible for the death he deals."[1] But there are many objections to this, each of which seems conclusive.[2] One is that if one can deliberate about whether to do an act, it cannot be true that, if one decides to do it, all responsibility for one's action will lie with someone else. But even if it were true that all responsibility for one's act would go to someone else, it would not follow that it would be permissible for one to do the act. It cannot become permissible to do an otherwise impermissible act just because the responsibility for it would go to someone else. At most, all that would follow from one's lack of responsibility is that one would be excused.

Defenders of the traditional theory of the just war have of course advanced other arguments to show that even in the most obviously unjust wars, fighting in accordance with the rules is always permissible. But these arguments have, in my

1 Augustine, *The City of God*, translated by Marcus Dods (New York: Modern Library, 1950), Book I, chapter 21, p. 27.

2 Jeff McMahan, "War," in David Estlund, ed., *The Oxford Handbook of Political Philosophy* (New York: Oxford University Press, 2012), pp. 308-9.

view, been no more successful than the two just cited.[3] So let us assume that it is morally impermissible to kill people solely because they are soldiers who impede the achievement of unjust aims in an unjust war. Consider, then, a soldier in a democratic society who has been commanded to fight in a war that is intended to achieve aims that are all objectively unjust. Suppose he knows that the war is unjust and that if he fights in it he will kill people who are merely attempting, by discriminate, necessary, and proportionate means, to defend themselves and others against his state's unjust aggression. Suppose he recognizes that it is impermissible for him to fight and on that ground refuses to do so. How ought his society to respond?

In contemporary democracies, selective conscientious objectors within the military are comparatively rare and are often dealt with informally – for example, by being assigned a role that does not involve combat or even close combat support. This is less troublesome than forcing such people into disobedience and then punishing them for it, which would inevitably provoke unwelcome controversy. But if a sufficient number of soldiers were to refuse on moral grounds to fight, it might not be feasible to reassign them all to non-combat functions. At some point, the military, and the government it serves, must implement a prearranged policy for dealing with conscientious objectors.

The central issue addressed by the essays in this book is whether a state may permissibly prohibit and penalize selective conscientious objection, particularly by active-duty members of the military, or whether it ought to provide legal options for conscientious objection, perhaps combined with certain non-punitive penalties imposed to deter exploitation of these options by those who seek to avoid combat for non-moral reasons.

One relevant consideration that seems uncontroversial is that if a war is objectively unjust and the state punishes a soldier who conscientiously refuses to fight in it, the state thereby wrongs that soldier. In fighting an unjust war, a state wrongs many people. It wrongs all those on the opposing side, both soldiers and civilians, who are harmed by the war fought against them. But it also wrongs its own soldiers. It wrongs them if it deceives them into fighting voluntarily and it wrongs them if it coerces them to fight through a threat of punishment for refusing to fight. It is obvious that they are wronged by being made to expose themselves to grave risks for the sake of goals that are unjust. And it is a further, distinctive wrong to be made, through deception or duress, to act in ways that are objectively immoral. So when a state's war is unjust, the state wrongs its own soldiers whether it succeeds in getting them to fight or fails to do so and then punishes them for disobedience.

The problem, however, is that while punishing selective conscientious objectors wrongs them when they are right that the war is unjust, it may not wrong them if they are mistaken and the war is in fact just. Yet a policy for dealing with selective conscientious objectors must operate in conditions of uncertainty and

3 For critical discussions of the other main arguments, see Jeff McMahan, *Killing in War* (Oxford: Clarendon Press, 2009), chapters 1 and 2.

disagreement, in which, at least at present, there is no impartial and authoritative source of judgment about whether a war is just or unjust.

One might say, then, that the punishment of selective conscientious objectors simply carries the same risk that is inherent in all forms of state punishment – namely, the risk of mistakes that will result in the punishment and wronging of innocent people. And, just as the aims achieved by punishment in general – social defense, deterrence of crime, and so on – outweigh the inevitable but relatively infrequent mistakes, so, one might argue, the values threatened by selective conscientious objection are sufficiently important to justify the inevitable mistakes made in the punishment of selective conscientious objectors. Central among these values is the ability to fight efficiently in a just war, particularly a just war of national self-defense. If soldiers were to have the legal right to opt out of particular wars, there would be greater uncertainty about their state's ability to fight effectively, even in a just war of self-defense. For some soldiers may sincerely believe that a war that is in fact just is unjust, while others may exploit the option of selective conscientious objection to evade, for non-moral reasons, their duty to fight. Furthermore, any uncertainty about the state's ability to fight would have the further effect of weakening its ability to deter aggression against it.

There are, however, relevant differences between the punishment of selective conscientious objection and the punishment of criminal action in general. Unlike most criminals, most people who claim to be selective conscientious objectors are acting on the basis of sincere moral convictions that, even when mistaken, are based on a generally beneficial reluctance to cause harm. There is also a major difference between the two cases in the ratio between instances of punishment that wrong the victim and those that do not. Because of the presumption of innocence and the burden of proof required for conviction, the percentage of convictions that result in the punishment of the innocent is relatively low. But the situation is different with the punishment of selective conscientious objectors. This is because there are at least as many unjust wars as there are just wars, and possibly more. For whenever there is a just war, it seems that the adversary against which it is fought must be fighting an unjust war. Moreover, given the highly plausible assumption that soldiers are more likely to object on moral grounds to participation in a war that is in fact unjust than to a war that is in fact just, it follows that any sincere conscientious objector is statistically more likely to be justified than to be unjustified. Hence the punishment of any particular selective conscientious objector is more likely than not to wrong him.

There are various other relevant differences between criminal punishment in general and the prohibition and punishment of selective conscientious objection but I will mention only one more. This is that while the deterrent effects of the punishment of ordinary crime are generally good, and socially desirable, the deterrent effects of punishing selective conscientious objection are mixed: some good, some bad. The good effect I have already mentioned: it is that there is a greater guarantee of military efficiency and effectiveness when the probability of any significant level of disobedience is low. This is highly important when a state needs to fight a just war. But there is an important bad effect that is less often noted, which is that soldiers are also deterred from refusing to participate in unjust wars, which makes it easier for governments to initiate such wars. If governments instead had to fear conscientious objection on a significant scale in response to the initiation of an unjust war, they would be less likely even to try. For a significant level of conscientious objection within a state's own military would be profoundly humiliating and delegitimizing to the government. If soldiers were to believe that participation in a war that lacks a just cause is morally wrong, and if there were a legal option of selective conscientious objection by active-duty soldiers, the effect would almost certainly be a decrease in the number of unjust wars, which would have as a corollary a decrease in the need to fight just wars in response.

Perhaps the most significant question, then, is which effect of having a legal option of selective conscientious objection would be greater: the inhibition of unjust wars or the weakening of the ability of states to fight just wars (and the consequent weakening of deterrence of unjust wars). It seems clear that an option of selective conscientious objection would be much more likely to be exercised in the case of a war that is in fact unjust than in a war that is just. This is in part because, as history shows, there is a strong bias toward believing that a war fought by one's own state must be just. This bias may be particularly strong among soldiers who have been commanded to fight. For there are usually many pressures operating to get them to fight and if they can foresee that these pressures are likely to succeed, they may seek, if only subconsciously, to shield themselves from guilt by allowing themselves to be convinced that the war is just. Because of this natural bias, it is quite easy for soldiers to believe that an unjust war in which they have been commanded to fight is just and quite difficult for them to believe that a just war is unjust. And, at least in the case of a just war of national self-defense, any reasons they might have to believe that it is unjust are likely to be opposed by considerations of self-interest. Thus, while soldiers who believe their state's war is just are in general at least as likely to be wrong as to be right, those who believe their state's war is unjust are in general more likely than not to be right.

This alone, however, is insufficient to settle the matter, for there is another disturbing complication. It is primarily liberal, democratic, and largely unaggressive states that are most likely to permit selective conscientious objection. States that are more aggressively disposed, particularly those that are undemocratic, are unlikely to be impressed by the arguments in favor of selective conscientious objection. But if the unaggressive states permit conscientious objection while the aggressive ones do not, that could give aggressors a systematic advantage over their victims.

It may not be enough to appeal again to the claim that soldiers are highly unlikely to mistake their own state's just war of national self-defense for an unjust war. But there is more that can be said. One point is that some democratic societies have already experimented with certain forms of conscientious objection and do not seem to have suffered any ill effects as a result. In the United States, for example, soldiers are not just permitted but legally required to disobey a "manifestly unlawful order" received in the course of a war. A hundred years ago this would have been unthinkable. The response to such an idea would have been that it would wholly undermine the chain of command and lead to paralysis on the battlefield. But that has not happened. Proposals for legally permitting selective conscientious objection by serving military personnel seek to extend this permission from the *jus in bello* to *jus ad bellum*. But they are significantly weaker, in that they propose that we permit, but not require, disobedience to a manifestly unlawful order to fight at all.

Another point is that the alternative to granting rights of selective conscientious objection is to force soldiers to bear an unfair share of the burden of national security. They are already asked to bear the physical risks of combat. And they bear the moral risk of inadvertently engaging in gravely immoral action, which in turn increases the risk of profound psychological damage if they survive the war. Meanwhile the civilian beneficiaries of their action, for all their pious expressions of "support for our brave men and women in uniform," do nothing to provide them with more impartial moral guidance than that which they receive from their government. It may seem intolerable if, in addition to this, the society punishes them when they have the courage and strength of character to refuse to do its bidding when it demands that they kill people as a means of achieving unjust aims. It may well be that a fair distribution of the burdens in society requires that civilians share some of the risks that go with respecting moral constraints on the pursuit of national security.

Introduction
'Sometime they'll give a war and nobody will come'[1]

Andrea Ellner, Paul Robinson, and David Whetham

In a lecture delivered in 1539, entitled *De Iure Belli* (*On the Law of War*), Francisco de Vitoria commented that:

> If the war seems patently unjust to the subject, he must not fight, even if he is ordered to do so by the prince. This is obvious, since one may not lawfully kill an innocent man on any authority, and in the case we are speaking of the enemy must be innocent. ... So even soldiers, if they fight in bad faith, are not excused. ... And from this flows the corollary that if their conscience tells subjects that the war is unjust, they must not go to war even if this conscience is wrong.[2]

Vitoria's view was that a soldier was not obliged to determine for himself whether a war was just or unjust; the soldier was quite entitled to leave that deliberation to his rulers. But if for some reason he happened to come to the conclusion that his rulers were wrong and the war was unjust, then he must refuse to fight. In this respect, he followed the advice of Thomas Aquinas that human beings must always follow their conscience, as long as the judgements of their conscience are well informed.[3]

In our enlightened age, and nearly half a millennium after Vitoria's careful considerations, there are almost no armies anywhere in the world that offer their soldiers this option. Most states do recognize that pacifists who object in principle to all wars have a right to refuse to fight in them. Even if in practice states often make life very difficult for those who try to exercise this right, conscientious objection of this absolute sort is relatively uncontroversial. By contrast, very few states are willing to allow people to refuse to fight if it is a particular conflict which they consider to be unjust. A soldier who tries to follow Vitoria's advice

1 Carl Sandburg, *The Complete Works of Carl Sandburg*, revised and expanded edition (New York: Harcourt Brace Jovanovich, 1969), p. 464.

2 Francisco de Vitoria, cited in Gregory M. Reichberg, Henrik Syse, and Endre Begby, *The Ethics of War: Classic and Contemporary Readings* (Oxford: Blackwell, 2006), pp. 318-19.

3 Thomas Aquinas, *Summa Theologiae* (London: Blackfriars, 1964), 1a2ae, 19, 5.

is liable to be punished for disobeying orders, desertion, or some other crime relating to disobedience or duty. Although there are some notable exceptions, such as Germany, most states adhere to a position similar to that adopted by the US Supreme Court, which in a landmark case in 1971 (Gillette v. United States) ruled that 'the exemption for those who oppose "participation in war in any form" applies to those who oppose participating in all war and not to those who object to participation in a particular war only.'[4]

Conscientious objection to specific wars rather than to wars in general is often referred to as 'selective conscientious objection'. Until recently, few people challenged the distinction between absolute and selective conscientious objection. However, in the past ten years, this has changed. Acts of selective conscientious objection in Israel during the intifada and in the United States and the United Kingdom during the war in Iraq have led some to reappraise the situation and argue that selective conscientious objection ought to be permitted. Whereas previously there had been little academic discussion of the topic, a critical mass of writing on the subject has recently emerged to bring it out of the darkness and into the forefront of ethical debate. This includes a special volume of the *Israel Law Review* in 2002,[5] a number of articles in the *Journal of Military Ethics*[6] and other academic journals,[7] and several book chapters which either explicitly support selective conscientious objection[8] or strongly imply support for it.[9] Together these provide the basis for carrying out a deeper and more thorough review of the topic.

Several factors lie behind the new interest in selective conscientious objection; some of these are political, some social, and some philosophical. On the political side, the war in Iraq, in particular, undermined one of the key arguments against selective conscientious objection, namely that soldiers should accept that their political masters know best and have information at their disposal which they do not, so they should accept the decisions of the politicians. Those who reluctantly swallowed their doubts about Iraqi weapons of mass destruction and then found that the doubts were actually very well-founded and that the politicians had seemingly misinformed them are now less inclined to accept this argument again.

4 *Gilette v. The United States*, 401 U.S. 437, 8 March 1971.

5 *Israel Law Review*, 36/3, (2002).

6 For example: Paul Robinson, 'Integrity and Selective Conscientious Objection', *Journal of Military Ethics*, 8/1 (2009), pp. 34-47; 'Saying No: Command Responsibility and the Ethics of Selective Conscientious Objection' Special Issue, *Journal of Military Ethics*, 8/2 (2009).

7 For example; George Clifford, 'Legalizing Selective Conscientious Objection', *Public Reason*, 3/1 (2011), pp. 22-38.

8 For example, J. Carl Ficarotta, 'Should Members of the Military Fight in Immoral Wars? A Case for Selective Conscientious Objection', in J. Carl Ficarotta, *Kantian Thinking About Military Ethics* (Farnham: Ashgate, 2010), pp. 73-86.

9 For example, Richard A. Gabriel, *The Warrior's Way: A Treatise on Military Ethics* (Kingston: Canadian Defence Academy Press, 2007), pp. 107-34.

On the social side, modern soldiers are more educated and better informed than their predecessors in generations past. The argument that they are in a state of 'invincible ignorance' and not in a position to make an informed decision seems less convincing given the ubiquitous nature of the contemporary media. At the same time, ethical education, which is taken seriously across all Western military training and educational institutions, focuses on building morally autonomous individuals, which seems incompatible with claims that the same soldiers must ignore their consciences regarding the justice of the wars in which they fight.

On the philosophical front, recent work by Jeff McMahan and David Rodin in particular has challenged some of the traditional philosophical foundations of the rejection of selective conscientious objection.[10] The traditional position has been that the ethics of whether it is just to wage war at all (*jus ad bellum*) and the ethics of what one is permitted to do during war (*jus in bello*) are related but still are also distinct from each other. It follows from this that having a just cause does not afford greater license to conduct that war beyond the rules of *in bello*, and a soldier can (and must) fight justly even in an unjust war. The larger justice of the war is not therefore his or her concern since it has no relevance to the morality of his or her actions or what he or she is permitted to do. McMahan and Rodin, however, have put forward powerful arguments suggesting that the *jus in bello* and *ad bellum* may be far more interdependent than previously assumed, and that therefore it is logically impossible to fight justly in an unjust war. If the war itself is wrong, killing someone in order to prosecute that injustice cannot be made acceptable through some arcane and arbitrary notion that morality is judged by different rules when applied to war. If this is so – and this is far from universally accepted – then refusing to fight in an unjust war becomes a moral obligation.

Together, these factors mean that the time is ripe for a fresh and thorough evaluation of the topic of selective conscientious objection. This collection of essays will examine the subject from a number of different perspectives, both theoretical and practical, looking both at the arguments for and against selective conscientious objection and at how different nations have dealt with, and continue to deal with, soldiers who refuse to fight in wars which they consider unjust.

Before analysing these arguments in more detail, we need first to define selective conscientious objection more precisely. The key word is 'conscientious'. Somebody who believes that a particular war is stupid, counter-productive, or ineffective policy does not qualify as a conscientious objector. He or she must believe that the war is morally wrong, and that participation in it would therefore be contrary to his or her conscience. The first does not necessarily imply the second: a soldier might consider a war unjust but still consider it his or her duty to participate due to other issues which are more important than the injustice of the war. He or she might, for instance, feel that the obligation not to abandon his

10 See for example the arguments in David Rodin and Henry Shue (eds), *Just and Unjust Warriors* (Oxford: Oxford University Press, 2008).

or her comrades outweighs the requirement not to fight. Selective conscientious objection thus refers to a fairly narrow set of cases.

As Asa Kasher has pointed out, determining what defines conscience in this context is difficult, as a result of which he concludes that introducing selective conscientious objection would be 'bound to involve conceptual tensions and inconsistencies'.[11] Following similar logic, some opponents of selective conscientious objection claim that most of those who claim such status do not really object to the wars in question on grounds of conscience but on grounds of politics, i.e. their objections are at their core political rather than moral. Such writers therefore draw a distinction between those who carry out conscientious objection and those who carry out civil disobedience and state that people who make a claim to belong in the first category in reality often belong in the second.[12]

For some, the common conflation of political and conscientious objection and the difficulties of distinguishing between the two are major reasons to maintain the prohibition against selective conscientious objection. Certainly, these issues raise serious practical questions about how one can determine whether somebody really is a conscientious objector. On the other hand, supporters of selective conscientious objection might point out that conscientious objection tribunals in many countries have been successfully making assessments of people's intent for decades, and the problem of determining whether somebody's objections are truly ones of conscience should in theory be no more difficult for a selective objector than for an absolute one.[13]

The primary distinction here is one of intent: the political objector aims through his civil disobedience to change policy; the selective conscientious objector aims to protect his own conscience. The latter may, of course, also hope to change policy, and may therefore have two motives, but the second one is what defines his objection as conscientious.[14] Issues of intent are not, however, the only ones. Also important are concerns about the consequences of permitting selective conscientious objection and about whether it is compatible with democratic values and more specifically with military ethics. The subject thus crosses the boundaries of deontological, consequentialist, and virtue ethics.

The case for selective conscientious objection rests in large part on Aquinas's assertion, mentioned at the start of this chapter, that a person should always obey his or her conscience, even if there is a possibility of being wrong. A soldier who cannot in good conscience fight in a given war would be acting immorally if he

11 Asa Kasher, 'Refusals: Neglected Aspects', in *Israel Law Review*, 36/3 (2002), p. 179.

12 In particular, see Avi Sagi and Ron Shapira, 'Civil Disobedience and Conscientious Objection', *Israel Law Review*, 36/3 (2002), pp. 181-218.

13 For instance, D. Malament, 'Selective Conscientious Objection and Gilette Decision', *Philosophy and Public Affairs*, 1/4 (1972), pp. 378 and 380.

14 For a discussion of these issues and a rebuttal of Sagi and Shapira, see: Alon Harel, 'Unconscionable Objection to Conscientious Objection: Notes on Sagi and Shapira', *Israel Law Review*, 36/3 (2002), pp. 219-26.

or she nevertheless chose to fight, whether it be due to fear of punishment or any other reason. Aquinas did, however, add an important proviso, which was that conscience needs to be informed. If you act according to your conscience and turn out to be mistaken, this is forgivable if you were in a state of 'invincible ignorance', such as when you have made a genuine effort to make an informed judgement but facts were hidden from you which caused you to make a mistake. In this case, it is not your fault that you were ignorant of the truth. If, however, you acted without proper consideration even though you could have discovered the truth if you had tried, you would be in a state of 'vincible ignorance'; your ignorance of the truth is your own fault. In this case, mistakes cannot be justified.

A crucial question in the debate about selective conscientious objection is whether soldiers exist in a state of invincible ignorance.[15] According to some, they do not; soldiers are quite capable of judging the justice of wars correctly. They are often very well educated, and most of the relevant information is in the public domain. If they choose to fight in an unjust war, they will therefore be in the realm of vincible ignorance, as they could have discovered the truth if they had but tried.

According to others, however, most soldiers are simply are not in a position to know all the relevant facts about a government's decision to go to war and so are not able to reach an informed judgement on the justice or injustice of the war. While in some contexts, the position taken by one's state might be clear enough, in others it might be horrendously complex and the ability to find out the truth about what is really going on may be compromised in many ways. In such circumstances, surely, they cannot be to blame for fighting in the war, only held accountable for their actions actually within the war if those actions violate the dictates of conscience directly, such as targeting non-combatants. If they have doubts about the justice of the actual war, they should exercise humility and accept that they should not follow the inner voice of their consciences but defer to those who are in a better position to judge. This might of course require a good deal of faith in one's chain of command – soldiers must have a genuine belief that the right questions have been asked by the right people at the top of that chain and that only once those senior officers have been satisfied with the answers will orders be issued to everyone else in the chain of command. This is exactly how the Chief of the General Staff in the UK, General Sir Peter Wall, views it:

> It is for the high command to ensure that intended operations satisfy *jus ad bellum* and to be prepared to reassure the members of the force that this is the case through the chain of command. To do otherwise would be a distraction of the highest order – potentially leading to chaos ...[16]

15 For discussions of invincible ignorance in this context, see: Andrew Sola, 'The Enlightened Grunt? Invincible Ignorance in the Just War Tradition', *Journal of Military Ethics*, 8/1 (2009), pp. 48-65.

16 Peter Wall, 'The Ethical and Legal Challenges of Operational Command', David Whetham (ed.) *Ethics, Law and Military Operations* (Basingstoke: Palgrave Macmillan, 2010), p. 222.

In this view, the responsibility is very clearly on the senior military leadership alone to examine the justice of the war and to try, through giving advice to the ruler, to avert a conflict if it is deemed unjust (even if those arguments are ultimately ignored or overruled).[17] Everyone else can rest assured that the appropriate questions have been asked. If such faith in their senior commanders is present, and it is indeed true that soldiers beneath this highest level of command are or can be invincibly ignorant, then they bear no responsibility for an unjust war in which they fight, only for their actual actions during a war. Thus *ius ad bellum* and *ius in bello* remain entirely distinct. For this reason, soldiers on both sides of the battle – just and unjust – can be justified in fighting and should be treated equally. From this follows the principle of the 'moral equality of combatants' – because of the lack of metaphysical certainty over who is objectively in the right or wrong, 'it cannot matter who started it'.[18]

The subject of selective conscientious objection is thus closely bound up with debates about the validity of the *ad bellum/in bello* divide and of the concept of the moral equality of combatants. If these ideas are valid, then a soldier need not worry about fighting in an unjust war as he or she bears no moral blame for it. The state, which does shoulder the responsibility, is then within its rights to oblige him or her to participate in the war.

However, as noted above, this traditional divide has come under real challenge from a number of angles. The moral equality of combatants in particular seems incompatible with modern ideas of human rights. The soldier fighting on the just side in a war has done nothing to justify another soldier's taking away his right to life. As Vitoria said, he is 'innocent'. If you are fighting on the unjust side there is no way that you can kill someone on the morally blameless side justly. It follows that *ius in bello* is intimately connected to *ius ad bellum* and that it is impossible to fight justly in an unjust war. If, therefore, you believe that the war is unjust, you should not participate in it.

Of course, this is all very well in theory, but the view runs into practical difficulties, since once one eliminates the divide between the two aspects of just war theory and eliminates also the traditional assumption of moral equality of combatants, one opens the door for everybody who considers himself or herself 'just' (in other words almost everybody) to consider that he or she is subject to different rules from his or her opponents, thereby undermining the laws which restrain behaviour in war. This could have seriously negative practical consequences. For this reason, Jeff McMahan concludes that even though the moral equality of combatants is not a morally valid concept, for legal reasons we need to preserve it.[19]

---

17 Whetham, 'The Just War Tradition: A Pragmatic Compromise' in *Ethics, Law and Military Operations*, p. 74. See also George R Lucas, 'Advice and Dissent: The Uniform Perspective', *Journal of Military* Ethics, 8/1 (2009).

18 Whetham, 'The Just War Tradition', p. 74.

19 Jeff McMahan, 'The Morality of War and the Law of War', in Rodin and Shue (eds), *Just and Unjust Warriors*, pp. 19-43.

This introduces an important distinction between morality and law. If the arguments against the *ad bellum/in bello* divide are correct, then from a moral point of view the soldier whose conscience tells him that the war is unjust should refuse to fight, but that does not necessarily mean that states should permit selective conscientious objection as a matter of law. In essence, the soldier must follow his conscience, but he or she should not complain about being punished for doing so. As John Rawls puts it:

> No doubt it is possible to imagine a legal system in which conscientious belief that the law is unjust is accepted as a defence for noncompliance. ... But as things are, such a scheme would presumably be unstable even in a state of near justice. We must pay a certain price to convince others that our actions have, in our carefully considered view, a sufficient moral basis in the political convictions of the community.[20]

In a military context, Colonel John Mattox comments that 'Army officers must recognize their responsibility ... to refuse an order ... [but] they must also possess the maturity to recognize, and the fortitude to accept, the personal consequences of refusal.'[21] Against this, however, Robinson observes that this view is 'a little perverse ... in that it is proposed to weed out the morally righteous by punishing them.'[22] David Enoch argues also that what he calls a 'willingness to pay the price' is in no way linked to the moral justification of refusal to fight, which is either right or wrong regardless of the willingness.[23] The sharp distinction between the morality and the law of war is not one which everybody accepts.[24]

McMahan's conclusion regarding this distinction draws on consequentialist reasoning. This is true also of many of the arguments against selective conscientious objection. States are essential to human flourishing – we need them to provide law, order, and security. But they are under continual threat and need to defend themselves. They therefore require armies and have to be able to rely on these armies whenever they need to use them. States could not carry out their basic task of protecting themselves and their citizens if they were uncertain whether their troops would fight. Selective conscientious objection would introduce such uncertainty. The higher social good therefore requires that it be prohibited by law. Following on from this, in the eyes of some, selective conscientious objection

20 John Rawls, *A Theory of Justice*, revised edition (Cambridge, MA: The Belknap Press, 1999), p. 322.

21 J.M. Mattox, 'The Ties that Bind: The Army Officer's Moral Obligations', in L.J. Matthews (ed.), *The Future of the Army Profession* (Boston: McGraw-Hill, 2002), p. 303.

22 Robinson, 'Integrity', p. 45.

23 David Enoch, 'Some Arguments Against Conscientious Objection and Civil Disobedience Refuted', *Israel Law Review*, 38/1 (2002), pp. 250-52.

24 For example, see Henry Shue, 'Do We Need a Morality of War?', in Rodin and Shue (eds), *Just and Unjust Warriors*, pp. 106-11.

is incompatible with democratic values and could have a damaging impact on democratic states. Democracies operate on the principle of majority rule: people agree to submit to collective decisions. If you allow people to opt out of any such decisions they do not like, you will undermine democracy. This is a particular concern in Israel, where there are fears of left-wing objectors refusing to fight against the Palestinians and right-wing objectors refusing orders to remove settlers from Palestinian land. Given the alleged confusion between political and conscientious objection, permitting the latter in such highly politically charged circumstances could have grave consequences and make it very difficult for the state to enact any policy effectively.[25]

The fear (or hope) here is that expressed by American poet Carl Sandburg, 'Sometime they'll give a war and nobody will come' in the title of this chapter.[26] For opponents of selective conscientious objection, this would seriously undermine states' ability to function. Supporters see it as perhaps a good thing: if sufficiently large numbers of people objected to a given war to make it difficult to fight, then that might be a very good sign that there was something wrong about the war, which the state ought to heed.[27]

Both opponents and supporters, however, face the same problem that confronts anybody making a consequentialist argument, namely that they do not actually know what the consequences of introducing selective conscientious objection would be. One can draw on evidence from general conscientious objection, which would suggest that the number of objectors is not sufficient to detract from states' ability to wage war, but since very few states have introduced selective conscientious objection, and even fewer have applied it during war time, we lack the evidence to draw firm conclusions. This is likely, therefore, to remain a highly disputed area.

Since agreement on consequentialist grounds is most unlikely, a more profitable avenue of debate is provided perhaps by virtue ethics. The issues here revolve around what behaviour is appropriate for soldiers given the specific role morality of the military profession and the virtues which that morality demands. Official lists of military virtues vary from country to country and also vary between services within countries. Most common are variations of loyalty or comradeship, courage, discipline or obedience, self-sacrifice, and integrity.[28] There are good

---

25 For a discussion of this issue, see: Barak Medina, 'Political Disobedience in the IDF: The Scope of the Legal Right of Soldiers to be Excused from Taking Part in Military Activities in the Occupied Territories', *Israel Law Review*, 38/1 (2002), pp. 73-110, especially pp. 91-4.

26 Sandburg, *Complete Works of Carl Sandburg*, p. 464.

27 Robinson, 'Integrity', pp. 43-4; Malament, 'Selective Conscientious Objection', p. 382; J.E. Capizzi, 'Selective Conscientious Objection in the United States', *Journal of Church and State*, 38/2 (1996), pp. 339-63.

28 Paul Robinson, 'Introduction: Ethics Education in the Military', in Paul Robinson, Nigel de Lee, and Don Carrick (eds), *Ethics Education in the Military* (Aldershot: Ashgate, 2008), pp. 5-7.

practical reasons for all of these. One does not, as a general rule, want soldiers who are endlessly questioning orders and putting their own interests and concerns before those of their comrades or the good of the mission. Selective conscientious objection runs deeply counter to many of these military virtues, which perhaps explains the visceral reaction that many military people have to the idea. On the surface it seems incompatible with the professional military ethic.

To take the example of comradeship, Asa Kasher comments that, 'Comradeship is an essential value of military ethics. … Acts of civil disobedience that takes the form of refusal by reserve officers and NCOs to serve within the framework of their combat units cause damage to the necessary sense of military comradeship. Thereby they inflict damage on the democratic regime, which owes its citizenry an effective military force of self-defence.'[29] The selective conscientious objector puts saving his own conscience above his responsibility to help his comrades. This is incompatible not only with the necessary virtue of comradeship but also with the virtue of self-sacrifice.

Similar arguments about the incompatibility of selective conscientious objection and the military professional ethic can be made with regard to other key virtues, most notably obedience and discipline. Obedience by soldiers to civil authority, and by junior soldiers to more senior ones, is arguably essential for the smooth running of any military force as well as for civil-military relations. The idea that soldiers can pick and choose which orders to obey based on their own consciences is anathema to the way that most people view the military.

Having said all that, it is perhaps unwise to view virtues as absolutes rather than as Aristotelian means. The key Aristotelian virtue of prudence or practical wisdom requires one to understand when a given virtue is appropriate and when it is not, and to what extent. Courage is a good thing, but not when it turns into recklessness. Obedience is virtuous, but not when it becomes submissiveness and obsequiousness. Loyalty is desirable, but not when it leads us to turn a blind eye to the faults of others. Among the military virtues listed above is integrity, which implies cohesion of belief and action. Arguably, a soldier cannot hold onto strong beliefs about the injustice of a war and yet submit to orders to fight in that war without destroying his integrity.[30] According to this line of argument, the professional military ethic demands that soldiers refuse to obey unethical orders. Thus Richard Gabriel comments that, 'Soldiers fail to live up to their oath to serve the nation if they do not speak out when they conclude that civilian authorities or military superiors are carrying out policies that they think are ethically wrong.'[31] This is closely linked to questions of individual autonomy. As Gabriel notes, 'It is a well-established principle of ethics that human beings cannot abandon their ethical autonomy and judgement to other human beings and thereby escape responsibility for their actions. So, too, soldiers as ethical beings and as members of an ethical

29 Kasher, 'Refusals', p. 177.

30 See Robinson, 'Integrity', for a detailed argument to this effect.

31 Gabriel, *The Warrior's Way*, p. 108.

profession cannot escape responsibility by acquiescing to unethical orders, even if the orders are legal and issued by a legitimate civilian or military authority.'[32]

Contemporary military ethics education in many cases aims deliberately at producing soldiers who are capable of autonomous moral decision-making. This fits with the perception that the conditions of modern warfare, in the era of the 'strategic corporal', mean that soldiers are no longer automata who have to do little more than stand in line and fire when told to, but will find themselves in difficult situations in which they have to decide *for themselves* what to do.[33] For supporters of selective conscientious objection, it seems inconsistent that military forces should attempt to make their members morally autonomous and then punish them for displaying that autonomy when they determine that the war in which they have been told to fight is morally wrong. Against this, though, opponents of selective conscientious objection might argue that the realm of soldiers' moral autonomy is limited to the narrow tactical sphere in which they operate and does not, and should not, extend to higher areas of policy, such as the decision whether to wage war.

The contributions to this volume explore and develop all these themes, starting with the theory and then moving onto the practice. Part I explores the arguments for and against selective conscience objection. To begin with, Brian Imiola, himself a serving Lieutenant Colonel in the US Army and professor at West Point, argues that in order to attempt to answer the question of whether or not to fight, soldiers must employ moral diligence and not simply default to claiming invincible ignorance. While one might question the ability of every 'common soldier' to employ moral diligence, some are certainly capable. Because of this, one might divide soldiers into different groups based on whether they are 'in the know', 'willingly ignorant', or truly 'can't know'. Different degrees of moral culpability for participating in an unjust war might logically follow from this categorization.

In the next chapter, Emmanuel Goffi explores the very notion of military obedience itself. He notes that the understandable obligation to obey orders in the military is deeply ingrained and this, unsurprisingly, leads to an apparent fit with deontological or duty-based moral reasoning such as that of Kant. However, Goffi warns us that this relationship is actually very superficial – the type of duty that Kant articulated was neither about passively obeying orders nor about subordination to arbitrary rules. Blindly following orders (or what Goffi calls 'low cost deontology') leaves no room for nuance and leads to intellectual poverty, since it gives no margin to military personnel to interpret those rules in the most appropriate manner. Both low cost deontology and its corollary, 'passive obedience', are fertile ground for irresponsibility and are therefore antithetical to what Kant would have understood as duty, which is actually about individual rationality and the ability of human beings to think autonomously about their duties. As such, this

32 Gabriel, *The Warrior's Way*, p. 109.

33 See Charles Krulak, 'The Strategic Corporal: Leadership in the Three Block War', *Marines Gazette*, January (1999).

understanding fits most comfortably with the Mission Command model used by both the British and US armed forces where subordinates are empowered to carry out their mission in the way that is most pertinent given the situation they are in, in accordance with the commander's overall intent, rather than by simply obeying specific orders given out regardless of local factors or events unfolding subsequent to those orders, although, of course, even the Mission Command model does not empower (or even permit) those same subordinates to question the overall intent or, indeed, the *ad bellum* justifications for the military action in the first place. Goffi argues that there is no actual moral duty to blindly obey any orders and that deontology, narrowly understood, is not only inappropriate when it comes to military behaviour, but is actually profoundly dangerous if applied without consideration for the likely consequences of one's actions.

In her chapter, Melissa Bergeron, explores selective conscientious objection starting with the military covenant between the state and its military. The implicit agreement is that the state (or its prince) is entitled to demand military service and in return, those who fight are absolved of responsibility for their actions – the responsibility for the war lies with the leader and not with those who simply act on behalf of the state. This is what underpins the assumption of the moral equality of combatants and also, Bergeron argues, forms the basis for many of the objections to embracing selective conscientious objection. To accept selective conscientious objection would be to damage the social contract and lead to an undermining of the democratic accountability for decisions to wage war. It is precisely the fact that those in military service who dissent to particular deployments are punished that gives their dissent moral value – something that would be lost should it just be one of the available options of normal service. While she finds the *status quo* to be morally obscene when one considers what is actually done by unjust combatants, she also argues that the assumption of the moral equality of combatants is, in practice, like having to accept that some guilty people will have to be allowed to go free to ensure that the many innocent do not suffer and are also permitted to remain free.

Michael Skerker proses a slightly different line of reasoning in an *Empirical Defence of Combatant Moral Equality* by challenging the recent trend towards the individualisation of moral reasoning and accountability within the military sphere. While the assassination of Caesar by a group of senators can be reduced to individual acts and individual as well as group culpability for those acts, the nature of contemporary military activity and the huge range of supporting tasks and roles required for someone to actually be involved in delivering lethal force is why all Western military institutions take on something akin to a corporate responsibility for all actions. As long as selective conscientious objection is not permitted and only legal orders are obeyed, this relieves individuals within those organisations from being ethically or morally culpable for participating in military activity. This formal equality of combatants becomes much more problematic if selective conscientious objection is allowed, but Skerker also presents a number of other practical reasons as to why it would be untenable if adopted as a policy by states.

In *Who Guards the Guards?* David Fisher provides a traditional-style defence of the importance of civilian masters maintaining objective control of the state's military. This objective control requires and maintains the very clear and strict division of moral responsibility between the *jus ad bellum* level of war, where political decisions about whether to go to war are determined and justified, and the *jus in bello* level, where military actions are actually taken to pursue those political goals in a military environment.

Written from the point of view of a soldier with multiple tours of duty under his belt, Dan Zupan concludes this section of the book with a chapter that reflects the tensions and challenges for the individual soldier of trying to apply all of the potentially conflicting requirements of loyalty and integrity along with duties to various different concepts and institutions. Zupan accepts that a curtailment of certain individual liberties is permitted due to the nature of the service offered when joining an all-volunteer force. The chapter refers to real life cases in which any requirement to investigate the moral basis for a short notice deployment would be hopelessly impractical for those hurriedly loading equipment onto military transporters, arguing that it is precisely the practical difficulties in ascertaining epistemic certainty in many cases that requires us to keep the *ad bellum/in bello* distinction and the corresponding division of moral responsibility. However, Zupan also argues that not all situations are like this and that even while accepting the traditional divide in moral responsibility in practice he can see 'no version of a social contract, agreed upon for the purpose of securing justice, that would bind me knowingly to participate in the mass murder that is unjust war'. A just society is therefore required to accept selective conscientious objection for those who are convinced that the war they are being ordered to participate in is unjust.

The case studies section of the book begins with an examination of the evolution of conscientious objection in Australian legal and political thought by Stephen and Nikki Coleman (with Richard Adams). Beginning with the 1903 Australian Defence Act, ground-breaking in its recognition of the right of a conscripted individual to be exempted from military service on the grounds of conscientious belief, it then looks at individual cases from the two World Wars, through periods of national service and the war in Vietnam, through to the 1991 Gulf War and beyond, charting how those cases were dealt with, and how the jurisprudence evolved as a result. The chapter concludes by referring to the difference in attitudes in law between those who are conscripted and those who volunteer, arguing that is morally objectionable that those who have volunteered to serve their country and to risk life and limb in the process, should be thought to have fewer rights than those who are compelled into the same service.

Stephen Deakin looks at the way that selective conscientious objection has historically been dealt with by the British military establishment and how a pragmatic approach has, at least until now, prevented it from becoming an overtly political issue. In cases where military law has not been broken in a blatant manner, the practical approach of simply discharging or moving those soldiers who raise objections of this type, means that those who are prepared to fight do not have to go to war alongside those who are not. This position is seen institutionally, implicitly if not explicitly, as being preferable to damaging corporate morale and cohesion.

Yossi Nehushtan then turns attention to the Israeli experience. Nehushtan argues that the distinction between conscientious objection and selective conscientious objection as understood by the state is actually meaningless from a moral point of view because one cannot choose one's deepest beliefs. Having explored the implications of this argument, the chapter then discusses some recent Israeli cases that demonstrate the 'misguided approach' of the Israeli Supreme Court when ruling on examples of political or selective conscientious objections.

Yves LeBouthillier's chapter examines the subject of this volume from a slightly different perspective by exploring how the Federal Court of Canada and the Federal Court of Appeal have dealt with applications for refugee status from those who have entered Canada from abroad claiming refugee protection by asserting selective conscientious objection to military service in their home state. The chapter notes that through the evolution of jurisprudence in this area, the courts have given primacy to the concept of 'state protection' and have proven unwilling to grant refugee status to those who cannot prove that state protection is lacking in their home country. Only once this hurdle has been passed will the Canadian courts consider whether they face persecution at home. This means that it is very difficult for a claimant from a democratic state to receive refugee status, although it is theoretically possible for those from non-democratic states.

The context of total defeat and the fall of the Nazi regime following World War II led to the adoption of a new German constitution in 1949. This 'Basic Law for the Federal Republic of Germany', guaranteeing all Germans certain fundamental human rights, is the backdrop to Jürgen Rose's chapter, which forms a bridge between the case studies section and the final section of the volume. German citizens have the right to freedom of faith and conscience as well as the right to refuse to perform military service. Unlike in many other states, this right does not become suspended just because someone joins the military and remains applicable to everyone, soldier or civilian, regular or reservist, conscript or volunteer. At the time of the introduction of this law, many feared that such an interpretation would forever neuter the ability of the German state to defend itself and its people. This chapter explores the practical implications of the German position and the implications for the relationship between the individual, the military and the state.

Building upon the preceding chapters, the final part of this volume looks at the practical advantages and disadvantages of implementing a policy of selective conscientious objection within the state and proposes some potential frameworks that might structure such responses. Carl Ficarotta proposes some guidelines for how a system of selective conscientious objection could be implemented if the principle itself were to be embraced by a state. He suggests that given the profound moral and psychological barriers that are present when one is considering disobeying a legal order, any system that would permit selective conscientious objection should, ideally, be designed to be as straightforward as possible, with a simple application procedure to avoid putting even more 'roadblocks' in the way. However, while he demonstrates that the difficulties of introducing and administering such a straightforward process would not be insurmountable, he also accepts that a more complex and difficult-to-use approach would probably be more acceptable to any state for political reasons – i.e. being realistic, even if a state did embrace selective conscientious objection, it is likely to want to keep as many roadblocks as possible to ensure it is a very difficult thing for any individual to claim this right. Despite conceding this, Ficarotta concludes that such an imperfect system would still be preferable to having no system at all.

Andrea Ellner rounds off this section by examining the role of war resisters in the US and Britain and the implications that this has for traditional assumptions about civil military relations, the military covenant and the practicality of selective conscientious objection. The chapter explores the experience of dissenters to the recent conflicts in Iraq and Afghanistan, their motivations, their actions and the results of these actions. Ellner demonstrates the profound moral courage shown by many of these resisters as they seek to uphold their moral beliefs regarding conceptions of justice or respect for human life in the face of substantial cost, both in professional and personal terms. Using this evidence drawn from recent experience, she challenges the traditional epistemic assumptions about the division of moral labour into those who can be expected to know enough to make informed decisions at the top of the chain of command and those who cannot at the bottom. Acknowledging that no-one in the chain of command should be considered to be a mere moral automaton would be a good step in positively reinforcing ethical behaviour, functioning as a bottom-up corrective for systemic problems that can result in breaches of ethical standards. When combined with the setting of clear ethical standards of behaviour from above, this approach would support a military culture that is likely to reduce ethically problematic orders and behaviour.

The concluding chapter then draws together and engages with the wide range of arguments presented in the volume and looks at the costs and benefits, military and societal, for both retaining the status quo and the possible implementation of different forms of selective conscientious objection. Referring to their 2010 editorial for the Journal of Military Ethics, Deane-Peter Baker proposes applying the 'Cook-Syse test' when writing on military ethics.[34] The purpose of such writing is not merely to produce 'wonderfully logically developed, conceptually clear, rigorously argued – and in the end professionally irrelevant' material, but instead to offer something that can also be of actual, 'real-world guidance for policy-makers, military commanders and leaders, or operational decision-making'.[35] The editors of this volume hope that through exploring this subject with an eye on the practical rather than just the theoretical implications, we have managed to shed some much-needed light on this complex subject and in the process, satisfy the

---

34 Deane-Peter Baker, 'Making Good Better: A Proposal for Teaching Ethics at the Service Academies', *Journal of Military Ethics*, 11/3 (2012), p. 210.

35 Martin L. Cook and Henrik Syse, 'What Should We Mean by 'Military Ethics'?', *Journal of Military Ethics*, 9/2 (2010), p. 120.

Cook-Syse test. The subject of selective conscientious objection is an issue of profound importance, both to the military and to the society that the military is supposed to serve. In an age of discretionary conflicts where the survival and even vital national interests of the state are rarely if ever at stake, the issue of when it is right and proper for individuals to take life on behalf of that political community is likely to become more problematic rather than less.

PART I
Arguments For and Against Accepting Selective Conscientious Objection

Chapter 1
The Duty of Diligence: Knowledge, Responsibility, and Selective Conscientious Objection

Brian Imiola

The view that soldiers are simply 'servants of the state' is popular among many soldiers, political leaders, and citizens. While the core identity of a soldier certainly includes servant of the state, we should not overlook the fact that soldiers are also individual, moral agents who are responsible for their actions. While soldiers may be fulfilling their contract with the state when called upon to deploy and fight the state's wars, their service to the state does not eliminate their moral responsibility to examine the moral justness of the wars they are being ordered to wage. In other words they must employ moral diligence in an effort to seek knowledge concerning the justness of the conflict. If soldiers find the war to be unjust, they have a responsibility to not participate in it and to become selective conscientious objectors.

Too often, soldiers choose not to employ moral diligence to the degree that is required when dealing with the seriousness of moral questions concerning war and killing. Instead they turn to one of several arguments in their defence. These include the argument that it is not a soldier's responsibility to seek knowledge to determine if the war is just or not – that responsibility rests with their political masters. Linked very closely with this argument is a second one focused on the impossibility of soldiers actually being able to obtain knowledge concerning the justness of the war. We can call this the 'invincible ignorance' argument. Its central claim is that if soldiers cannot determine the justness of the war then their participation in it can be morally excused. A final argument returns to the primacy of a soldier's identity as 'servant of the state'. It claims that the oath soldiers took to serve the state is the decisive factor requiring soldiers to override moral concerns and obligations regarding participation in an unjust war in favour of their state imposed duties.

My position is that soldiers must employ moral diligence to attempt to obtain knowledge concerning any conflict they are asked to participate in. In supporting this view, I hope to demonstrate that the arguments in favour of soldiers either dismissing knowledge concerning the justness of war, being unable to obtain such knowledge, or overriding the moral obligation that such knowledge implies are flawed. Determinations of both responsibility and decisions to serve in war or become a selective conscientious objector both ultimately rely on seeking and achieving knowledge of the war's justness. Finally, senior military leaders have a special responsibility to seek knowledge of the war's justness as their actions and decisions will influence their subordinates' decisions on whether to participate in a conflict.

Invincible Ignorance, the Traditional Division of Responsibility within the Just War Tradition

An excellent starting point for a discussion of responsibility and knowledge can be found in Shakespeare's *Henry V.* In Act 4, Scene 1, a disguised King Henry walks through his camp to judge his army's morale prior to the battle of Agincourt. He declares to one group of soldiers that 'Methinks I could not die anywhere so contented as in the King's company, his cause being just and quarrel honourable.' One soldier responds with 'That's more than we know.' A second soldier adds 'Ay, or more than we should seek after, for we know enough if we know we are the King's subjects. If his cause be wrong, our obedience to the King wipes the crime out of us.'[1]

Henry's conversation with the two common soldiers mirrors the view found in the Just War Tradition and its separation into two parts: *jus ad bellum* – the justice of the war – and *jus in bello* – justice in the war. The first part concerns itself with the justifications for resorting to force as a means of achieving political objectives. *Jus in bello* concerns itself with the considerations on how a war is prosecuted. Historically, the Just War Tradition holds that while soldiers may be held responsible for their own misconduct in fighting the war (the *jus in bello* portion), they are not responsible for *jus ad bellum* considerations which are the responsibility of the state's political leaders.

Shifting momentarily from the moral realm to the legal realm, this means that soldiers are generally held responsible for only two of the three classes of war-related crimes formulated by the International Law Commission at the request of the UN General Assembly in 1950. These three crimes are: Crimes Against Peace which include (a) planning, preparation, initiation or waging of a war of aggression or a war in violation of international treaties, agreements or assurances and (b) participation in a common plan or conspiracy for the accomplishment of any of the acts mentioned under (a); War Crimes which are violations of the laws or customs of war, which include, but are not limited to, murder, ill-treatment of the civilian population, murder or ill-treatment of prisoners of war, killing of hostages, plunder of public or private property, wanton destruction of cities, or villages, or devastation not justified by military necessity; and Crimes against

1 Anthony E. Hartle, 'Moral Principles and Moral Reasoning', in Don M. Snider and Lloyd J. Matthews (eds), *Forging the Warrior's Character: Moral Precepts from the Cadet Prayer* (Sisters, OR: Jerico, LLC, 2007).

Humanity which consist of murder, extermination, enslavement, deportation, and other inhuman acts done against any civilian population, or persecutions on political, racial, or religious grounds, when such acts are done or such persecutions are carried out in execution of or done in connection with any crimes against the peace or any war crime.[2]

The traditional view therefore holds that soldiers are not responsible for crimes against peace because the decision to go to war is a political decision rather than a military one. Since soldiers play no role in this decision they, in Paul Christopher's words, 'can never be responsible for the crime of war, qua soldiers.'[3] This reasoning results in a very neat division of *jus ad bellum* and *jus in bello* responsibilities between soldiers and their political masters. Initially, the traditional view seems reasonable. In the age of mass armies and conscription, can we really assign guilt and responsibility for crimes against peace to the common soldier and professional officer corps? Practically speaking, this would mean that, assuming the unjust side lost, its soldiers would be viewed as guilty and presumably subject to some kind of punishment. In general, punishment has not occurred at the conclusion of wars in the 19th, 20th, and 21st centuries. Wehrmacht soldiers after the Second World War and Iraqi soldiers after the Gulf War were not viewed as guilty for the crime of war nor punished for fighting for an unjust cause.

Assigning collective guilt to a military organization for crimes against peace seems problematic. As we have already seen, Christopher argues that soldiers can never be responsible for the crime of war. He not only focuses on the distinction between political and military decisions but also on the beliefs and knowledge (or lack of it) that soldiers possess. Here he appeals to both Michael Walzer and Francisco Vitoria. He interprets Walzer to say that in regards to soldiers 'theirs is never the crime of aggression because they always believe that their cause is just.'[4] In other words, soldiers cannot be held guilty when the knowledge they have leads them to believe their actions are just.

In reference to Vitoria, Christopher appeals to the notion of invincible ignorance. Ignorance is 'lack of knowledge about a thing in a being capable of knowing'. Ignorance categorized as invincible occurs 'when a person is unable to rid himself of it.' Ignorance which is invincible thus provides a 'valid excuse and excludes sin.'[5] Christopher holds that soldiers are always infected with invincible ignorance when it concerns the justness of the war they are being called to fight in. Presumably, Christopher sees this as a desirable state. He even goes so far as to ask 'Who could wish it otherwise?' However, his own answer to this is not couched in

2 Donald A. Wells (ed.), *An Encyclopedia of War and Ethics* (Westport, CT: Greenwood, 1996), p. 362.

3 Paul Christopher, *The Ethics of War and Peace: An Introduction to Legal and Moral Issues* (Englewood Cliffs, NJ: Prentice Hall, 1994), p. 96.

4 Christopher, *The Ethics of War and Peace*, p. 97.

5 'Ignorance', in *The Catholic Encyclopedia*, http://www.newadvent.org/cathen/07648a.htm.

terms of knowledge or knowing but in terms of duty. He argues that 'It would be a traitorous abrogation of public trust to swear an oath of preparedness to perform certain duties on demand in return for "meals and support by the state", and then, after accepting the benefits of the agreement, to refuse to fulfil the duties stipulated when called upon to do so. No citizen logically could wish it so.'[6] While no citizen could logically wish a soldier to shirk his duties when called upon to fight a just war, it is not at all clear that a citizen would want a soldier to actively support an unjust war. A simple analogy shows this to be the case.

Reconsidering Responsibility

Consider the case of the bystander who sees a criminal rob a bank and begin to load sacks of money into the get-away car. Since the sacks of money are heavy and there are quite a few, the robber asks the bystander for help. The bystander willingly complies and assists in the loading of the get-away car. Upon finishing the loading, the criminal drives away.

In this case, it seems that our bystander is responsible and guilty of wrongdoing. Assuming the bystander had knowledge that a criminal act was being committed, recognized a moral wrong, and then proceeded to assist in the act's completion, it seems clear we would label this person as an accomplice to a crime and as someone who has acted immorally.

The same labelling holds true for soldiers who agree to fight in a war which they know is unjust and thus morally wrong. While one might debate what is the best course of action for the bystander to pursue in the bank robbery scenario (call the police, run for help, attempt to stop the bank robber), one should not condone their actively aiding the robber. If this is true, there is no reason we should expect anything less from a soldier in similar circumstances. We may argue about the appropriate actions for the soldier to take (refuse to fight, apply for selective conscientious objection status, or resign his or her commission) but no one should expect a soldier to actively support a war which he or she knows to be unjust and to involve unjustified killing.

In fact, Vitoria makes it quite clear that soldiers should not blindly march off to war. In *Political Writings*, Vitoria lays out the following four propositions:

> First, if the war seems patently unjust to the subject, he must not fight even if he is ordered to do so by the prince. ... Second ..., all those who are admitted or called or of their own accord attend the public or royal council are in duty bound to examine the cause of just war. ... Third, lesser subjects who are not invited to be heard in the councils ... are not required to examine the causes of war, but may lawfully go to war trusting the judgment of their superiors. ... Fourth, there may nevertheless be arguments and proofs of the injustice of the

6 Christopher, *The Ethics of War and Peace*, p. 97.

> war so powerful, that even citizens and subjects of the lower class may not use ignorance as an excuse for serving as soldiers.[7]

The first, second, and fourth propositions are clear in that they require moral examinations of the causes of war by all of those concerned. The third proposition is somewhat problematic but still understandable within the overall framework of Vitoria. This proposition is of a practical nature. Vitoria discusses the impossibility and inexpediency of putting the arguments of 'difficult public business before every member of the common people'. This proposition can also be seen to rely on the second proposition in that the lesser subjects (the majority of soldiers) must rely on trusting their superiors who are duty bound to examine the cause of the war (we shall examine this in more detail in a later section). The first and fourth propositions also allow lesser subjects to opt out of the war if they have knowledge of its injustice.

What we see at work in Vitoria's propositions is the role of 'moral diligence'. Ignorance is said to be truly invincible only when a person is unable to rid himself of it notwithstanding an attempt to do so. The requisite for this diligence 'must be commensurate with the importance of the affair in hand, and with the capacity of the agent, in a word such as a really sensible and prudent person would use under the circumstances.'[8] With this in mind, and considering the importance of an affair such as war making or fighting, it is clear that soldiers must examine the justness of the war that they face and not simply throw up their hands and claim invincible ignorance.

Knowledge and Responsibility

Whether soldiers can be held responsible for fighting in an unjust war rests on the knowledge they have or can acquire regarding the justness of the conflict. Likewise the decision to be a conscientious objector relies on this same knowledge. Assigning blame to those who knowingly and willingly participate in unjust acts seems reasonable enough. Similarly, it seems reasonable not to expect someone to knowingly engage in unjust acts. However, what about those who make little or no effort to determine if the acts they are planning on participating in are unjust? It thus seems that we have three categories of knowledge/ignorance in regards to participating in an unjust war. These categories are those who have knowledge that the war is unjust, those in a state of vincible ignorance regarding the justness of the war, and those who are truly in a state of invincible ignorance.

---

7 Francisco de Vitoria, 'On the Law of War', in Gregory M. Reichberg, Henrik Syse, and Endre Begby (eds), *The Ethics of War: Classic and Contemporary Readings* (Oxford: Blackwell Publishing, 2006), p. 319.

8 'Ignorance' in *The Catholic Encyclopedia*, http://www.newadvent.org/cathen/07648a.htm.

For the first category, consider the case of General A, who sits in on his political master's explanation for going to war. He learns that, despite appeals to *jus ad bellum* principles the real reason for going to war is to jump start the economy and to divert the nation's focus from the severe internal economic problems it is experiencing. It is clear to General A that his state is about to embark on a war of aggression. If he supports the war, he is no different than the bystander who recognizing the robber was committing a crime by stealing from the bank, nevertheless went on to assist him. Like the bystander, General A is an accomplice to a crime, specifically the crime of aggression (crimes against peace). General A is also morally responsible for knowingly participating in an unjust war.

In his opening statement in the case against those Germans being tried as war criminals, Robert H. Jackson made it clear that at certain times soldiers can and should be held responsible for the crime of war or crimes against peace. As Jackson put it: 'We recognize that to plan warfare is the business of professional soldiers in every country. But it is one thing to plan strategic moves in the event war comes, and it is another to plot and intrigue to bring that war on.'[9] Jackson went on to say that the leaders of the German High command were not on trial 'because they lost the war but because they started it.'[10]

The Nuremberg Tribunal stated in regards to certain high ranking military leaders of the German high command that there was 'much evidence as to the participation of these officers in planning and waging aggressive war ... without their military guidance the aggressive ambitions of Hitler and his fellow Nazis would have been academic and sterile.'[11] Because of their knowledge that the war they were embarking upon was unjust and their subsequent efforts to plan and facilitate aggression, the Tribunal recommended that those who were guilty of these crimes should not escape punishment.[12]

Since the conclusion of the Nuremberg trials, international law has expanded liability in regards to crimes against peace. Article 25 (3) (d) of the International Criminal Court (ICC) now provides that where a group of persons are acting with a common purpose, they incur liability if the offence is committed or attempted, where their contribution is intentional and is made with the aim of furthering the criminal activity or criminal purpose of the group, where such activity or purpose involves the commission of a crime or made in the knowledge of the intention of the group to commit the crime.[13] Clearly by this standard, soldiers who have knowledge

---

9 Robert H. Jackson, *The Case Against the Nazi War Criminals*, (New York: Alfred A. Knopt, Inc., 1946), p. 85.

10 Jackson, *The Case*, p. 85.

11 Jan Philipp Reemtsma, 'On War Crimes', in Omer Bartov, Atina Grossmann and Mary Nolan (eds), *Crimes of War: Guilt and Denial in the Twentieth Century* (New York: New Press, 2002), p. 12.

12 Reemtsma, 'On War Crimes', p. 12.

13 *Rome Statute of the International Criminal Court*, Article 25 (3) (d), http://untreaty.un.org/cod/icc/statute/romefra.htm.

of such activity can and should be held both morally and legally responsible. They have knowledge that the act they are planning or embarking upon is unjust.

While the case against those who have knowledge that their actions are unjust seems fairly clear, the case of those possessing vincible ignorance (that which can be overcome) is not so straightforward. Consider General B who, while not possessing knowledge of the fact that the war is unjust, has the capability to determine if the war is just. Imagine he has access to certain intelligence files and the minutes of key meetings and briefings. With the employment of moral diligence, General B could find out the truth of the situation. He need only access the files in question. However, he does not do so. Perhaps General B is reluctant to discover the truth that the war is unjust because he would feel obligated to take some kind of action (end his career by resigning, buck the chain of command, refuse orders, etc.). Unlike the person who knowingly assisted a criminal act, General B is not strictly liable for the crime. However, he has certainly not exercised moral diligence. Wilful ignorance of this kind might be labelled crass or supine.[14] At a minimum we must consider General B morally 'irresponsible' in that he did not take the steps we would expect of a prudent and reasonable man. We must also question the character of a person such as General B and to some degree find it lacking.

Although both examples given have used high ranking officers, the possibility does exist that due to their duty assignment, both junior and mid-level officers as well as enlisted soldiers could end up in the same situation as our imaginary generals. The examples focused on senior leaders because much of the key intelligence considered vital in determining whether to go to war may be compartmentalized at the higher levels of command in which political decisions are discussed.

The Professional Military Ethic and Individual Moral Responsibility

To this point, I have demonstrated that soldiers, like all moral agents, have a responsibility to determine the morality of actions they plan to carry out. Additionally, I have demonstrated that soldiers often can have knowledge about the justness of wars they are called upon to fight. It seems reasonable that if a moral agent determines that an action which he or she is asked to carry out is immoral, he or she should refuse to carry it out. However, many soldiers would argue that the moral obligation of their oath trumps other moral concerns including the duty not to harm others unjustly. In fact, the very nature of the military seems to condition soldiers to go to war without attempting to ascertain the war's actual 'justness'. One of the imperatives of military organizations is that they must be sufficiently disciplined with a recognized hierarchy of authority to ensure that orders are carried out consistently and reliably.[15] As such, certain traits and virtues

14 *Rome Statute*.

15 Hartle, 'Moral Principles and Moral Reasoning', p. 213.

are taught and internalized among professional militaries. Samuel Huntington's observation of the military profession makes this point clearly. He argues that 'For the profession to perform its function, each level within it must be able to command the instantaneous and loyal obedience of subordinate levels. Without those relationships, military professionalism is impossible. Consequently, loyalty and obedience are the highest military virtues.'[16]

An example of this focus on loyalty and obedience is particularly clear in the oath American soldiers take when enlisting:

> I do solemnly swear (or affirm) that I will support and defend the Constitution of the United States against all enemies, foreign and domestic; that I will bear true faith and allegiance to the same; and that I will obey the orders of the President of the United States and the orders of the officers appointed over me, according to regulations and the Uniform Code of Military Justice. So help me God.

In addition to loyalty and obedience to the state, another important aspect of the professional military ethic is loyalty to one's fellow soldiers. One sociologist argues this is actually the highest loyalty because while 'in theory the most important of these loyalties is to the United States Constitution; in practice the most important – to a soldier's morale and to his or her willingness to obey orders and assume responsibility – is to comrades.'[17]

This duty and loyalty to fellow soldiers is a cornerstone of the professional military ethic and its importance cannot be emphasized enough. The special relationship between military comrades with its emphasis on not letting fellow soldiers down is another decisive factor for going to war regardless of the war's justice.

The weight of the professional military ethic and oaths should not be downplayed. However, adherence to an ethic or oath should not be viewed as the ultimate 'trump card' requiring a soldier to fight in an unjust war. An oath, no matter how solemn, cannot require someone to carry out a moral wrong. For example, consider the case of a son who is called to the death bed of his father. They clasp hands and the father extracts a solemn oath from his son to carry out his last wishes as laid out in a sealed envelope that the father presses into his son's hand. The son agrees. A few moments later the father passes away. When the son opens the envelope, he is horrified to discover that his father's last wishes include hiring a hit man to kill some of his former neighbours, providing money to a terrorist organization, and requiring the son to kill his sister who had previously fallen out with the father. Given what the son is being asked to do, it is hard to see how the son could be morally obligated to carry out his father's wishes. While he took the oath in good faith, his reasonable expectation was that the requests he

16 Samuel, Huntington, *The Soldier and the State*, (New York: Vintage Books, 1964), p. 73.

17 Kim Hays, *Practicing Virtues: Moral Traditions at Quaker and Military Boarding Schools*, (Berkeley, CA: University of California Press, 1994), p. 54.

would be asked to fulfil would not be immoral ones. Such is the case with soldiers who take oaths. They should expect to follow through on their oaths to include fighting on behalf of the state in support of just causes such as self-defence or defence of others. However, there is no reasonable expectation that their oaths will require them to knowingly and willingly engage in unjustified harm to others.

Dealing With True Invincible Ignorance

If the decision to either participate in a war or refuse to do so rests on knowledge of whether or not the war is just, what should a soldier do when he or she lacks the ability to obtain that knowledge? To answer this concern, we must return to our definition of ignorance as a lack of knowledge about a thing in a being capable of knowing. 'Fundamentally speaking and with regard to a given object, ignorance is the outcome of the limitations of our intellect or of the obscurity of the matter itself.'[18] Given that the vast majority of soldiers are intellectually capable of understanding and applying *jus ad bellum* criteria, our focus will be on the 'obscurity of the matter', in other words the uncertainty that exists in determining if a war is just or unjust.

Hugo Grotius, a key figure in the development of the modern Just War Tradition captures the idea of 'obscurity' in a passage from Aristotle's *Nicomachean Ethics* which states that 'certainty is not to be found in moral questions in the same degree as in the mathematical sciences', and then goes on to quote Andronicus of Rhodes who said: 'It is hard to distinguish what is truly just from that which appears to be.'[19]

An example which illustrates the uncertainty concerning moral judgments in regards to the justness of wars is the public discussion concerning the Iraq War. In the period leading up to the actual invasion of Iraq by the United States and its allies in 2003, world opinion was split on what course of action was truly just. Even within America, there was a severe division over whether or not the war could be justified. Applying the same *jus ad bellum* principles led many to very different conclusions as summarized below.[20]

The Iraq War: Just or Unjust War?

1. The war should be declared on proper authority.
   Opponents: Only a new vote by the Security Council would grant proper authority.

---

18 'Ignorance'.

19 Hugo Grotius, 'De iure belli', in Gregory M. Reichberg, Henrik Syse and Endre Begby (eds), *The Ethics of War: Classic and Contemporary Readings* (Oxford: Blackwell Publishing, 2006), p. 413.

20 Rodney P. Carlisle, *Iraq War* (New York: Facts on File, Inc., 2007), pp. 66-7.

Proponents: Iraq had disobeyed prior Security Council resolutions that threatened force if disobeyed, and Resolution #1441, passed in 2002, gave sufficient authority.

2. The war should have a just cause.
   Opponents: Iraq had not attacked anyone and there was no clear evidence of weapons of mass destruction (WMDs), according to chief UN inspector Hans Blix.
   Proponents: There was no clear evidence that Iraq had eliminated its known WMDs, according to Blix. These WMDs were a threat to other states.
3. There should be a reasonable chance of success.
   Opponents: The casualty rate in a war might be so high that success would not be clear.
   Proponents: The Iraqi regime was weaker than it had been in 1991, and few troops would be needed to defeat the Iraqis.
4. The act of war should be in proportion with the offence of the nation being attacked.
   Opponents: Obstruction of inspections is not an act of war.
   Proponents: The regime was dangerous, had WMDs, and was a threat to security unless removed.
5. War should only be used as a last resort.
   Opponents: Inspections should continue before war is the only option.
   Proponents: Saddam Hussein had used up his chances to cooperate.
6. The real reason for the war should show right intent and should be the stated reason.
   Opponents: Prime Minister Tony Blair and President George W. Bush really wanted access to Iraqi oil or had other motives.
   Proponents: Removing WMDs from a ruthless, terror-sponsoring state was the real and sufficient reason.
7. The outcome of the war will lead to a more just result than if no war had occurred.
   Opponents: Thousands of civilian and military casualties could be avoided.
   Proponents: If the war was conducted quickly and humanely, the result will be freedom for the Iraqi people and the elimination of a serious threat to the region.

Besides the obscurity that sometimes surrounds moral issues, another reason soldiers might also be unable to rid themselves of ignorance despite their efforts at moral diligence can be the kind of regime or state they serve. One can easily imagine members of North Korea's Armed Forces, having been subject to propagandized education and indoctrination for their entire life, as incapable of being able to determine if a war is just or unjust. Their lack of access to information makes such a determination very difficult if not impossible. On the other hand, one can just as easily imagine a soldier in an open democratic society facing the same problem of uncertainty for different reasons. Instead of too little information, there is too much. Various opinions and interpretations in regards to the war exist within society, the media, and the world community. The best efforts of moral diligence result only in continued uncertainty. A state of invincible ignorance exists.

It is certainly not the case that a state of invincible ignorance is inevitable for the vast majority of soldiers. At times, it might be the case that the war one's state is about to embark on is clearly just or unjust. If it is unjust, soldiers are morally obligated to take action to avoid participating in the war. However in those cases where invincible ignorance does exist, the question remains of what soldiers should do.

There are various views within the Just War Tradition concerning what soldiers should do when they are truly in a state of invincible ignorance. For example, Grotius bases his view on the maxim 'what you question, do not do.' He argued that if there is doubt about an important matter, the safer course of action should be chosen 'because of the great difference between the courses to be chosen, the safer is to be preferred. ... Whence it follows that in case of doubt we must refrain from war ... Now war is of the utmost importance, seeing that in consequence of war most sufferings usually fall also upon innocent persons. Therefore in the midst of divergent opinions we must lean towards peace.'[21] For Grotius, those possessing invincible ignorance and a degree of uncertainty, should 'play it safe' and not participate in a war which may be unjust. The consequences of participating in a potentially unjust war are too serious to take chances on.

As we have already seen Vitoria disagrees with Grotius' 'play it safe' approach. He holds that soldiers in a state of invincible ignorance 'may lawfully go to war trusting the judgment of their superiors.'[22] Of these two views, it is Vitoria's which most soldiers will be drawn to as it fits with the military ethic, that in turn forms the core of the professional military culture. Given a state of invincible ignorance in which the justness of a particular war is unable to be determined, the soldiers' adherence to a professional military ethic centred around obedience, loyalty, duty, and a sworn oath are decisive factors in favour of going to war on behalf of the state, as long as the state and its leaders have proved trustworthy in the past.

Trust and the Responsibility of Senior Leaders

Our discussion of deciding whether to participate in a war has focused on the role of knowledge. In keeping with this focus, senior military leaders have a special responsibility as they generally possess more knowledge regarding the war and its justness. In addition to this knowledge, they also have the trust of their subordinates. A key component of the professional military ethic is trust between military professionals and within the chain of command. When told that the unit on their right will attack to support them and that another unit will conduct a

21 Grotius, 'De iure belli', p. 419.

22 Vitoria, 'On the Law of War', p. 319.

diversionary attack to distract the enemy from their own attack, soldiers trust the chain of command to make those things happen. At a more personal level, soldiers trust their leaders to make wise decisions, act morally, and, given the nature of their profession, do their best to protect them from harm. This trust is deeply ingrained.

The combination of knowledge and trust is what makes the role of senior leaders so important. In many cases, soldiers may not have the capability to access the knowledge necessary for them to determine if the war they are embarking on is just. In such a case, their professional military ethic will generally lead them to go off to fight. Obedience, loyalty, and trust point them in this direction. If senior leaders have knowledge that the cause is unjust based on the greater information they possess, they must take action. Whether this action takes the form of speaking out, publicly resigning their commission, or refusing to participate in the war's planning or execution, they are obligated to act. Not doing so is morally wrong in that it betrays the trust placed in them and will lead their subordinates to participate in an immoral activity.

On 5 February 2003, I was in a van travelling from Fort Hood, Texas, to the Austin Airport. Along with other officers, we were headed to Kuwait to conduct final planning for our part in the invasion of Iraq. Several of the officers had expressed private concerns over the justness of the possible war. As we drove, we listened to Secretary of State (and former Chairman of the Joint Chiefs of Staff) Powell's radio speech to the United Nations. Powell made a clear case for the justness of our cause and the evidence for it. Based on his position, former service, and the respect and trust in which he was held, Powell's comments and support for the invasion served to convince us (and at least a portion of the American people) that what we were doing was right. While I have no reason to doubt Secretary of State Powell's sincere belief that the information he was providing was correct, one can only imagine what the consequences would be if someone like Powell, the Chairman of the Joint Chiefs, or a Service Chief, having information that the war was unjust, was to speak out publicly. Such an action might even avert an unjust war from occurring.

Conclusion

It is not the case, as put forth by King Henry's common soldier that 'obedience to the King wipes the crime out of us.' Faced with a war that is clearly unjust, we would be morally wrong to aid or participate in it. Likewise, when soldiers enter the military, they are not issued a blanket of invincible ignorance to relieve them of the responsibility to seek the truth concerning the justness of a war they are called upon to fight. It is the responsibility of every soldier to exercise moral diligence to try and ascertain whether a war is just or not. In those cases in which the war is unjust, soldiers should understand that becoming a selective conscientious objector is not an abrogation of the public trust but instead a moral obligation.

Chapter 2

There is No Real Moral Obligation to Obey Orders: Escaping from 'Low Cost Deontology'

Emmanuel R. Goffi[1]

> What characterizes the free spirit is not that his opinions are the more correct but that he has liberated himself from tradition, whether the outcome has been successful or a failure.[2]
>
> Friedrich Nietzsche

In recent years, there have been many cases of selective conscientious objection, particularly in the United States and Israel. These cases must be considered in the context of more general debates about the morality of modern wars and of renewed reflections on just war theory,[3] in which some scholars have tried to determine the extent to which soldiers are individually responsible for unjust wars.[4] It seems from these reflections that selective conscientious objection is so controversial because many armed forces retain a strong tradition of obedience. According to this tradition, soldiers are expected to obey even if orders impose on them an obligation to participate in an unjust war. Using the example of the French armed forces, this chapter will demonstrate this tradition is mistaken, because there is in fact no moral obligation to obey orders.

The complexity of modern conflicts has made it increasingly difficult for soldiers to abide by the Law of Armed Conflict.[5] Consequently, one can observe a

---

1 The views expressed in this chapter are solely those of the author. They do not necessarily reflect the opinions of the French Ministry of Defense, the French Air Force, or even the institutions the author works for.

2 Friedrich Nietzsche, *Human All Too Human. A Book for Free Spirits*, translated by R.J. Hollingdale (New York: Cambridge University Press, 2005), p. 108.

3 For further and insightful developments on that point see Michael Walzer, *Just and Unjust Wars: A Moral Argument with Historical Illustrations* (New York: Basic 1977), pp. 309-16.

4 David Rodin and Henry Shue (eds.), *Just and Unjust Warriors: The Moral and Legal Status of Soldiers* (Oxford: Oxford University Press, 2008).

5 Emmanuel Goffi, *Les armées françaises face à la morale. Une réflexion au cœur des conflits modernes* (Paris: L'Harmattan, 2011), pp. 86-95.

'strengthening of the demand of ethics'[6] and a strong trend towards moral rules. In France this trend gave birth to so-called moral codes such as the *Code d'honneur du Légionnaire* in the 1980s,[7] and lately to the *Code du soldat*.[8] These display a strong propensity to deontology inasmuch as soldiers are obliged to obey rules and orders. This obedience can be observed in two forms. The first consists of obedience to civilian authority for, as Pierre Barral writes, 'subordination to the government's command appears in democracies as a principle of the legally constituted state (*État de droit*).'[9] The second one is the inner custom of military obedience to authority *per se*. This twofold subordination of military personnel to military and civilian authorities is mainly due to a very superficial, and undoubtedly comfortable way of understanding Kantian deontology as rules imposed from without which must be followed. However both are a product of the same logic of subordination to authority. This is the *raison d'être* of the French moral codes: obedience to authority whatever the authority. The result of these moral and legal norms is that obedience is a non-derogable imperative. Deontology deeply permeates the entire French military apparatus from both a legal and a moral point of view.

Yet, looking at the matter more closely, one notices that the French military operates a very superficial version of deontology, which I will refer to as 'low cost deontology'. This produces several results, among which are a sclerosis of the mind and a tendency to passive obedience, neither of which, as we will see, is supported by Kantian deontology. For deontology is neither about passively obeying orders nor about subordination to arbitrary rules. Deontology, as Kant puts it in the *Groundwork for the Metaphysics of Morals*, is about rationality and the ability of human beings to think autonomously about their duties. Furthermore, ethics is not limited to deontology. Virtue ethics and utilitarianism can and must be considered in any serious analysis of the moral obligation to obey orders. But neither in fact postulates any obligation to obey orders.

In this chapter I will argue through the example of the French armed forces that 'low cost deontology' is superficial and to some degree dangerous for it deprives military personnel of their moral autonomy, restricts thinking, and leads to passive obedience to subjective and partial rules and orders. Furthermore, I suggest that low cost deontology and its corollary 'passive obedience' are fertile ground for irresponsibility and that the French military should move from low cost deontology to a wider approach of moral reasoning which encompasses utilitarianism, in order to put responsibility at the core of the profession of arms.

---

6 Bastien Irondelle, 'L'armée française et l'éthique dans les conflits de post-guerre froide', *Critique Internationale* 41 (2008), pp. 119-36.

7 See http://www.legion-etrangere.com/modules/info_seul.php?id=92&block=1&titre=code-d-honneur-du-legionnaire.

8 See http://www.defense.gouv.fr/terre/bloc-les-essentiels/code-du-soldat, accessed 28 January 2013.

9 Pierre Barral, *Pouvoir civil et commandement militaire. Du roi connétable aux leaders du 20ᵉ siècle* (Paris: Les Presses de Sciences Po, 2005), pp. 186-93.

Deontology needs thus to be revisited, rethought, and deepened. Obviously, one also must draw a distinction between legal and moral obligations to obey orders. However, for the purpose of this chapter, I will focus on morality and put legality aside. In this framework I will mainly work from the *Code du soldat*, a supposedly 'moral code' in force in the French Army.

This chapter aims to show that, regardless of legal, strategic, or political considerations, there is no moral duty to obey orders, and that deontology, if misinterpreted or oversimplified, can turn to be irrelevant, if not dangerous, as a framework for determining military action.

The Moral Obligation to Obey Orders in France: The Overuse of 'Low Cost Deontology'

In 1964 Marshal Alphonse Juin wrote a monograph entitled *Three Centuries of Military Obedience (1650-1963)*[10] in which he calls for a return to the tradition of obedience, stressing that a distinction needs to be made between 'obedience within the forces and loyalty of the forces to those who lead the business of the State'.[11] This distinction, confirmed by the French *Statut Général des Militaires*,[12] can be considered irrelevant to some degree, because the distinction between military and political power is at a certain level quite blurred and both refer to obedience to authority, whatever it is. Indeed, this three-century tradition of obedience has even deeper roots in the long Christian history of France. The initial requirement of obedience to spiritual power, stated for instance in Deuteronomy,[13] gently turned into an obligation to obey temporal power.[14]

Later on General de Gaulle called for the rise of the 'Man of character' (*l'Homme de caractère*) able to decide on his own, avoiding passive obedience to impersonal rules too general to be efficient in particular cases.[15] Notably he stressed that 'the fact that he [*the man of character*] finds his power of decision within himself, and not imposed upon him by order, often disinclines him to adopt an attitude of passive

10 Alphonse Juin (Maréchal), *Trois siècles d'obéissance militaire (1650-1963)* (Sceaux: L'esprit du Livre, 2009).

11 Juin, *Trois siècles*, p. 16. Author's translation.

12 Loi n° 2005-270 du 24 mars 2005 portant Statut Général des Militaires, http://www.legifrance.gouv.fr/affichTexte.do?cidTexte=JORFTEXT000000808186&dateTexte=&categorieLien=id, accessed 25 September 2013. See more particularly Art. 1 and 8. The former calls for loyalty toward the nation, the latter for obedience to superiors' orders.

13 Deuteronomy 13:1: 'All this word which I command you, that shall ye observe to do; thou shalt not add thereto, nor diminish from it.'

14 For more on this shift see Ernst Kantorowicz, 'Pro Patria Mori in Medieval Political Thought', *The American Historical Review*, 56/1 (1951), pp. 472-92; or Ivan Strenski, *Contesting Sacrifice: Religion, Nationalism, and Social Thought in France* (Chicago: The University of Chicago Press, 2002).

15 Charles de Gaulle, *Le Fil de l'Épée* (Paris: Perrin, 2010), pp. 69-118.

obedience.'[16] In saying that, General de Gaulle was not appealing for disobedience, but rather for active obedience, demanding both reflection and an assumption of responsibility for the consequences of the decision thus taken.

It seems this call has been obscured since then. Looking at the preface of Marshal Juin's monograph, one can read General Irastorza, then Chief of the Army Staff, writing that 'obedience remains constitutive of the soldier and this notion must be understood, accepted, and internalized by each citizen who has made the choice to exercise the profession of arms.'[17] Several other normative texts similarly demand obedience to authority. An example is a 2005 instruction implementing the decree on general military discipline, which states: 'obedience to orders is the subordinate's first duty.'[18]

Many texts and regulations insist on obedience without deepening the notion. Some documents, however, such as *L'exercice du métier des armes dans l'armée de terre: fondements et principes* (The Profession of Arms in the Army: Foundations and Principles), issued by the Army Staff, do stress that obedience must not be passive.[19] At the end of the day, therefore, things seem unclear. One point remains constant, however: French soldiers must obey orders, i.e. rules imposed from the outside. This is shown in the traditional statement made during a change of command: 'On behalf of the President of the Republic, you will from now on recognize as your chief Colonel (or Captain) … here with us, and you will obey him in all he will command you, for the sake of the duty, the implementation of military regulations, the observance of laws and the success of France's arms.'

It is important now to say that the tradition of obedience highlighted by Marshal Juin has slipped from moral requirements made by religion to legal constraints imposed by laws. It is obvious that such regulations are essential for the military. There is no reason to dispute the usefulness of these rules in terms of military efficiency. The moral obligation of obedience, by contrast, is far more disputable.

Let us use the *Code du soldat* to examine this point further. This code, widely presented as a code of ethics, was written in 1999 by French General Jean-René Bachelet. Presented as a series of eleven short articles, the *Code du soldat* aims to provide soldiers with a framework of action. A first observation would be that there is nothing new in it. All the articles can be found in normative texts elsewhere. One could, however, say that at least soldiers now have a digest of rules which is easily transportable and can be read fast. This leads us to a second issue, namely that such short rules are unable to encompass all situations and they give soldiers the wrong impression that by following the *Code*'s rules they will act morally. From this arises a third problem, certainly the most dangerous: thinking that the

16 Ibid., p. 78.

17 Juin, *Trois siècles*, p. 9.

18 Instruction n° 201710/DEF/SGA/DFP/FM1 du 4 novembre 2005 d'application du décret relatif à la discipline générale militaire, article 7.

19 Etat-major de l'armée de terre, *L'exercice du métier des armes dans l'armée de terre : fondements et principes* (Paris, Janvier 1999), chapter 2, p. 12.

Code provides them with rules to act morally, soldiers do not take time to think about their behaviour by themselves.

Going back to obedience, the *Code* states in its fourth point that the soldier 'obeys orders in compliance with laws and international conventions'. Aside from the fact that obedience is already a legal obligation, this rule leaves no room for nuances. It follows that soldiers will just obey orders checking that they match the law. Since most military personnel do not know the laws, and most of the time they do not have time to look for them, one can easily conclude that they will obey passively, thus perfectly following the *Code*'s imperative. This is what I call 'low cost deontology'.

In short, such low cost deontology is made for dummies and produces a sclerosis of the mind. Furthermore, this sclerosis, rooted in obedience, diminishes responsibility. In this system, it would, after all, be legitimate for a soldier who obeys legal orders but whose actions cause immoral consequences, to consider himself or herself not responsible since he or she followed a code prescribing obedience to orders. As Hannah Arendt tells us, nobody can rightfully escape his or her responsibility by arguing he or she was just 'a cog' within the system.[20] Nevertheless, the argument above has been widely used throughout history, as it is appealing to those who follow the *Code*'s articles and are trained to strictly obey orders. This is dangerous.

By way of illustration, imagine that we are on 17 June 1940, listening to Marshal Pétain's address to the French nation saying: 'It is with a heavy heart that I tell you today that we must stop the fight.'[21] In accordance with the *Code*, should one obey this order? The following day, General De Gaulle, speaking from London, called the French to join him in continuing the fight against Germany. Should one choose this call over that of Pétain's? Both Pétain and de Gaulle were legitimate authorities invested with a right to be obeyed. The former was appointed Prime Minister of France by French President Albert Lebrun and the Parliament, and thus had a rational-legal authority. He also benefited from a charismatic authority he gained during the First World War. The latter was also considered as a charismatic authority due to his role during the Second World War.[22] In answer to these questions, the *Code* would imply that one should obey the orders given by Marshal Pétain, since the same code states that, 'Serving France, the soldier is entirely devoted to it, at any time and place.' At that time, Marshal Pétain officially represented France. Under the *Code*, therefore, General de Gaulle and his followers were morally condemnable. Yet this conclusion is highly disputable from both a deontological and a consequentialist point of view.

---

20 Hannah Arendt, 'Personal Responsibility Under Dictatorship', in Jerome Khon (ed.), *Responsibility and Judgment* (New York: Schocken Books, 2003).

21 'C'est le cœur serré que je vous dis aujourd'hui qu'il faut cesser le combat.'

22 Legitimate authority is here understood as Max Weber defined it in *The Three Types of Legitimate Rules*. See Reinhard Bendix, *Max Weber: an Intellectual Portrait* (University of California Press, 1977).

The example calls into question the very relevance of a 'code of ethics' based on low cost deontology. As General de la Maisonneuve writes: 'the soldier can never be told enough he must imagine that another "June 40" could ... confront him again with tragic situations.'[23]

Low cost deontology leads to many difficulties. It demands, through a kind of a categorical imperative, that a rule, once established, must be obeyed. This idea produces intellectual poverty since it gives no margin to military personnel to interpret rules or to decide by themselves what is morally relevant. Consequently, low cost deontology favours irresponsibility. Once a rule has been stated, it is quite easy to consider that the responsibility for deviant behaviour is the sole responsibility of the rulers.

There is undoubtedly a strong need, if not an obligation, for obedience in the military. But this obligation is not moral. It is legal. General de la Maisonneuve grasps the issue perfectly when he writes: 'certainly, submission to rules, implementation of orders, are the soldier's alphabet; but with letters one must be able to compose.'[24] By 'compose' he means nothing more than thinking about what one is doing. In other words, General de la Maisonneuve calls upon reason and rationality. In that he follows Immanuel Kant, and in the process indirectly tells us that it would indeed be entirely legitimate to refuse to fight in a war with which one does not agree.

What Does Kant Tell Us About Obedience?

Neither Kant's deontology, nor Bentham's utilitarianism, nor Aristotle's virtue ethics, impose a duty to passively obey orders. So what do they tell us? Since we stated that French moral codes are based on deontology, it seems useful to start with Kant's stance on moral obligation to obey orders.

Kantian deontology is anything but easy. It is not the purpose of this chapter to cover the full spectrum of Kant's thought on the matter. However, even without diving into the depths of deontology, it is clear that a lot of work has to be done in the French forces to strengthen its approach to ethics.

As mentioned earlier, the *Code du soldat* states in its fourth article that the soldier 'obeys orders in compliance with laws and international conventions'. This article, imposed upon the soldier from outside, is the expression of the soldier's duty. Here 'duty' is associated 'with legal requirements, requirements of one's role ... or one's office ... requirements of one's job'.[25] But Kant's understanding

23 Eric de la Maisonneuve, *Le métier de soldat* (Paris: Economica, 2002), p. 16. Author's translation.

24 de la Maisonneuve, *Le métier de soldat*, p. 15.

25 Marcia Baron, 'Acting from Duty', in Allen W. Wood (ed.), *Groundwork for the Metaphysics of Morals. Immanuel Kant* (New Haven CT, and London: Yale University Press, 2002), pp. 94-5.

of duty is quite different from mere social expectations. For him 'duty' is the outcome of reason, of the 'good will' which is itself part of duty.[26]

Furthermore Kant claims that 'there is nothing it is possible to think of anywhere in the world, or indeed anything at all outside it, that can be held to be good without limitation, excepting only a good will.'[27] Consequently the moral worth of a particular action lies in the will of the one who performed the action, not in the action performed, for it was imposed upon him as a duty. Duty is thus for Kant the outcome of one's good will not mere application of rules.

In the case of the *Code* it is worth stressing two points. First, if there is a need for such a code this means that soldiers have not internalized the articles of the *Code* through the reading of legal documents in which they already appear. Otherwise, they would obey without being told to do so. Thus, one can conclude that the articles are not the expression of the soldier's good will. After all, if one were deeply convinced that being altruistic is good, one would not need to keep a code in one's pocket saying 'be altruistic'. Following Kant's logic, if someone acts contingent on moral rules imposed upon him, the action is not considered as moral.

Kant makes an insightful distinction between acting *in conformity with* duty and acting *from* duty.[28] In the first case actions are commanded by external requirements, in the second case they are commanded by internal convictions. Consequently a soldier who obeys orders because it is made compulsory by a code does not act morally. Conversely, a soldier who obeys orders because he is convinced that it is a good thing does act morally. Therefore, if one is asked to fight an unjust war, one is not required to do so. To act morally one must be deeply convinced that fighting this war is consistent with one's own moral convictions.

In addition, one should keep in mind that 'an action from duty has its moral worth … on the *principle of the volition* … without regard to the ends that can be effected through such action.'[29] If a code makes obedience compulsory, it is legitimate to consider that any violation of the code would lead to sanction. Therefore soldier's obedience could be both mere submission without reasoning and the fruit of the fear of sanction. In the latter case, the logic is teleological, i.e., directed towards the potential consequences of the action, and thus of no moral worth. Following Kant's view of it, acting in conformity with the *Code du soldat*'s rules would therefore be immoral.

From a deontological point of view, what is expected is that soldiers act in compliance with maxims, i.e. 'the subjective principle of the volition'.[30] According to Kant, rational beings are free and benefit from autonomy of will so that they can make their own choices, even refusing to fight a particular war, without

26 Immanuel Kant, 'Groundwork for the Metaphysics of Morals', in Wood (ed.), *Groundwork*, p. 13.

27 Kant, 'Groundwork', p. 9.

28 Kant, 'Groundwork', pp. 13-14.

29 Kant, 'Groundwork', pp. 15-16.

30 Kant, 'Groundwork', p. 16.

obeying rules imposed from outside. Rationality is here understood as 'the ability to evaluate beliefs and acts (and so forth) with an eye to whether they meet the relevant standards, and to alter our beliefs and acts in light of those evaluations', as Shelly Kagan puts it.[31] This means that one should be not only able to evaluate acts and beliefs, but also to decide not to comply with them. In short, a soldier can step back and refuse to obey.

Put simply, the ability to evaluate the relevance of an order comes under the concept of autonomy of the will. Only actions made with autonomy are morally permissible. And so, according to Kant, '*morality* is thus the relation of actions to the autonomy of the will … That action which can subsist with the autonomy of the will is *permitted*; that which does not agree with it is *impermissible*.'[32]

Immanuel Kant was not the only one to develop this idea. Rousseau developed it in his *Social Contract*. According to the French philosopher, 'moral liberty' alone makes man 'truly master of himself',[33] and 'to remove all liberty from his will is to remove all morality from his acts.'[34] In his way Rousseau inspired Kant to think of human dignity and of the centrality of freedom.

There is no doubt that the soldier's code has been written without consulting French soldiers. Thus, its articles are not the expression of the will of each but an assertion of the authority of the hierarchy. As a result, from a moral point of view, there is no reason why soldiers should obey the *Code*.

According to Kant, the perfect expression of rationality and autonomous will is what he calls the *categorical imperative*, stated as follows: 'act only in accordance with that maxim through which you can at the same time will that it become a universal law.'[35] At the core of this imperative is the will. An act will be moral if, and only if, it is done willingly from duty itself, resulting from a categorical imperative prescribed autonomously by a rational being to himself.

One question needs to be addressed here: does the *Code du soldat* pass the formula of universal law? In his categorical imperative Kant suggests that a maxim to be obeyed should be universalized.[36]

Since the soldier's code is not a legal document, nobody is legally bound to apply it. The only potential obligation is moral. Now, if the *Code* is morally to be obeyed it must meet the requirement of Kant's deontology. Thus one would say that there is a categorical imperative saying something like, 'Always obey the *Code du soldat*.' But as we have seen before, this imperative must pass the

31 Shelly Kagan, 'Kantianism for Consequentialists', in Wood (ed.), *Groundwork*, p. 113.

32 Kant, 'Groundwork', p. 57.

33 Jean-Jacques Rousseau. *The Social Contract or Principles of Political Rights*, translated by G.D.H. Cole. Book I: 8, pp. 14-15 of the online version available at http://www.ucc.ie/social_policy/Rousseau_contrat-social.pdf.

34 Rousseau, *The Social Contract*, Book I: 4, p. 7.

35 Kant, 'Groundwork', p. 37.

36 Kant, 'Groundwork', p. 49.

universalization test. It is obvious that there is no reason at all that all humankind would agree to apply such a law. It is, furthermore, highly disputable that even the French would apply it. Even within the French forces this idea would be, and already is, disputed. As a result, we see that there is no moral obligation, from a deontological angle, to obey the *Code*.

Besides, even if obedience to the *Code* could pass the universalization test, it is doubtful that its articles would pass it as well. We have already seen that the articles of the *Code* cannot be morally acceptable since they violate the autonomy of the will of soldiers. Consider now that one universalizes the following article: 'he obeys orders in compliance with laws and international conventions'. History shows us that obedience, as well as compliance with laws, is far from universal. General de Gaulle did not obey orders coming from Marshal Pétain and his government, who in turn did not respect international conventions. As General de la Maisonneuve argues, 'all in the soldier's training must be a preparation to the worst, including being able to disobey.'[37] As far as the soldier's code is concerned, it would also be legitimate to ask if a law stating 'always obey passively orders' would be acceptable either.

As we have seen up to now, there is no moral obligation of obedience explicitly stated in Kantian deontology. On the contrary, Kant definitely calls for freedom of mind and autonomous will. Even the categorical imperative requires this freedom. This is far from an endorsement of articles set up at the top of the hierarchy and imposed upon soldiers without any consultation, and without asking them if they want to comply with these articles.

I must admit here that this is only the visible part of the iceberg. Kant's deontology is too complex to be dealt with here in its entirety. The aim so far has been merely to give a slight idea of how far the *Code du soldat* is from being a moral code, and thus how far low cost deontology, as applied in the French armed forces, is from deontology.

Obedience and Utilitarianism

In France deontology finds many more supporters than in some other countries. Indeed, in Anglo-Saxon countries, even if Kantian deontology is widely studied, utilitarianism is important also. So, if deontology does not support any moral obligation to obey orders, what about Bentham and Mill's utilitarianism and Aristotle's virtue ethics?

Ethics can be used to determine social conventions about good/right and bad/wrong which state what one must or must not do. These rules, though, do not encompass the full spectrum of ethics. Deontology is not sufficient in itself to decide whether an act is morally acceptable. In 1725, Francis Hutcheson wrote:

37 de la Maisonneuve, *Le métier de soldat*, p. 14. Author's translation.

> In comparing the moral Qualitys of Actions, in order to regulate our Election among various Actions propos'd, or to find which of them has the greatest moral Excellency, we are led by our moral Sense of Virtue to judge thus; that in equal Degrees of Happiness, expected to proceed from the Action, the Virtue is in proportion to the Number of Persons to whom the Happiness shall extend; ... so that, that Action is best, which procures the greatest Happiness for the greatest numbers; and that, worst, which, in like manner, occasions Misery.[38]

Thus, according to Hutcheson, the moral worth of an act is not to be found in the good will of the one who commits it, but in the outcomes of the act, and virtue proceeds from the principle of the greatest happiness. In other words the ends determine the morality of the act.

This idea gave birth to utilitarianism, subsequently deepened in the work of Jeremy Bentham and John Stuart Mill. Utilitarianism is a branch of consequentialism, focusing on the consequences of an act to decide whether the act in question was right or wrong. As Mill writes, 'the creed which accepts as the foundation of morals, Utility, or the Greatest Happiness Principle, holds that actions are right in proportion as they tend to promote happiness, wrong as they tend to produce the reverse of happiness.'[39] As far as obedience is concerned, for utilitarians the moral worth of every action lies in the consequences of such obedience and not in the fact that the author respected any given rule. This argument is used by consequentialists to reject any idea of selective conscientious objection on the basis that 'if it were allowed, the number of objectors would greatly increase and the ability of the state to wage war would be adversely affected.'[40]

Consequently, if one obeys an order following a rule like the one in the French soldier's code, whether it be *from* or *in conformity with* duty, and if that order results in bad consequences, the action will be considered as morally condemnable. It is worth stressing however that the consequences resulting from obedience to orders could also be worse than those resulting from disobedience. Put simply: an action is right if, and only if, it results in as much good as any available alternative.

All actions aim at reaching a particular end. According to Bentham, the only end worth being pursued is pleasure.[41] As Epicurus writes, 'we recognize pleasure as the first and innate good, and from this we originate every choice and aversion, and we return to this as the measure of feeling for judging every good.'[42]

---

38 Francis Hutcheson, *Inquiry into the Origins of our Ideas of Beauty and Virtue, in Two Treatises* (Indianapolis: Liberty Fund, 2004), Sect. 3.8, p. 125.

39 John S. Mill (1863), *Utilitarianism* (Kitchener, Ontario: Batoche Books Limited, 2001), chapter 2, p. 10.

40 Paul Robinson, 'Integrity and Selective Conscientious Objection', *Journal of Military Ethics*, 8/1 (2009), p. 43.

41 Jeremy Bentham, *An Introduction to the Principles of Morals and Legislation* (Kitchener, Ontario: Batoche Books Limited, 2000), I:I, p. 14.

42 Epicurus' Letter to Menoeceus, http://san.beck.org/Epicurus.html

So, is there a moral obligation to obey orders from a utilitarian point of view? Nothing is less certain! For a utilitarian soldier, obeying the *Code* or obeying orders more generally would require him to think about the consequences of his obedience. If the foreseen consequences of his obedience are worse than those of his disobedience, he should disobey regardless of what the *Code* says. The soldier should aim at maximizing the welfare of the greatest number, not only among his fellow citizens, but among 'all sentient beings' as Mill put it. Thus, although General de Gaulle could be morally criticized for leaving France and continuing fighting from the point of view of low cost deontology, from a utilitarian angle his decision was moral, since it undoubtedly resulted in the 'greatest good for the greatest number' compared to other available options.

It seems, therefore, that one is not morally compelled to abide by rules claiming an obligation to obey orders. Yet, it is not that easy. Bentham's act-utilitarianism postulates that the moral worth of an act is evaluated through the good it results in, but Mill also developed a kind of rule-utilitarianism which states that an act is morally permissible if it is required by a rule from which the maximization of welfare would ensue. One can then ask if a rule like 'always obey orders', or any rule like it, would be morally acceptable. To answer such a question one would have to ask if this could meet the greater happiness principle.

There is no reason at all, however, to suppose that all orders would result in the greatest good for the greatest number. The My Lai massacre is certainly a good example of this.[43] Many examples could be found in France as well. There is no need to cite the many misdeeds committed on orders in Algeria. Recently in 2005, Firmin Mahé's case[44] reopened the question of obedience to orders.

Any rule-utilitarianism would imply that the actor reflects on the consequences of his act. This means that one must not obey 'passively' but must take all the time needed to evaluate the foreseeable consequences of one's obedience. Pleasure, or happiness, being the intrinsic good, any action needs to be morally evaluated regarding all the criteria that help to measure it. Given the number of criteria, such as duration, intensity, fruitfulness, likelihood, quantity and quality of pleasure, the number of 'sentient beings' affected, different alternatives, and so on, any action would require rigorous assessment before one took a decision. This obviously disqualifies passive obedience.

Mill emphasizes qualities such as knowledge, intelligence, freedom and autonomy in his utilitarianism, claiming that 'over himself, over his own body and mind, the individual is sovereign.'[45] Utilitarianism being democratic, Mill fears that individuals could fall under the yoke of society, in which 'wherever there is an ascendant class, a large portion of the morality of the country emanates from its class interests, and its feelings of class superiority.'[46] This can be applied

43 Walzer, *Just and Unjust Wars*, pp. 309-16.

44 Goffi, *Les armées françaises*, pp. 78 and 99.

45 Mill, *Utilitarianism*, Ch. 1, p. 13.

46 Mill, *Utilitarianism*, Ch. 1, p. 10.

to the military. That is why, given that 'loyalty and obedience are the highest military virtues', and the capacity of human beings generally speaking to submit to authority,[47] 'when the military man receives a legal order from an authorized superior, he does not hesitate, he does not substitute his own views; he obeys instantly.'[48] The risk is that any moral consideration might be thus put aside.

In the *Code* the second article states: 'He [the soldier] accomplishes his mission with the will to win and vanquish, if necessary at the risk of his life.' Given that soldiers are expected to obey orders they also are expected to accept to give their life 'if necessary'. For utilitarians 'if necessary' would mean 'in order to reduce suffering' of all sentient beings. For the *Code* it likely signifies 'for the sake of the nation', as formally stated in the *Code de la défense* (Code of Defence).[49] According to the principle of utility, the morally best/better action is the one that produces the greatest/greater utility, i.e. the greater/greatest amount of happiness, to the greatest/greater number of people. We have also said that this would require deep scrutiny of the criteria which help to measure happiness. Amongst those criteria one seems worth stressing when it comes to thinking about sacrifice, namely fruitfulness: pain today may be the only way to avoid greater pain later on. Suppose one complies with the rule that one must obey orders. One would also comply with the rule saying that one must accept giving one's life 'if necessary', i.e. on a superior's order for reasons that are beyond one's understanding. But there is nothing *prima facie* which proves the fruitfulness of one's sacrifice. Looking at Afghanistan, for instance, one could query whether the sacrifices of soldiers, whatever their nationality, were fruitful. From a utilitarian point of view one does not have to accept giving one's life without knowing exactly what the consequences of such an act would be and without being sure it would result in the greater happiness for the greater number. It naturally follows that soldiers can refuse to participate in a war in which they would risk their lives for no clear reasons, or for the wrong reasons, as was the case in Iraq when the American and British administrations stated that the intervention was decided on the grounds that Iraq possessed weapons of mass destruction.

Obedience and Virtue Ethics

Utilitarianism is teleological in its orientation, focusing on the goal of actions and their expected consequences. But it never claims any obligation to obey orders

47 Stanley Milgram, *Obedience to Authority* (New York: Perennial Classics, 2009).

48 Samuel Huntington, The Soldier and the State. The Theory and Politics of Civil-Military Relations (Cambridge: Belknap, 1981), p. 73.

49 Article L4111-1, du *Code de la défense*, http://www.legifrance.gouv.fr/affichCodeArticle.do?cidTexte=LEGITEXT000006071307&idArticle=LEGIARTI000006540239&dateTexte=&categorieLien=cid, accessed 28 January 2013.

imposed upon oneself from outside. On the contrary it calls for complex reasoning aiming at a hedonistic assessment of the consequences of action.

Aristotle's virtue ethics is teleological and hedonist as well. Rather than the maximization of happiness, the main goals of virtue ethics are living well (*eudaimonia*) in communities and achieving excellence. In order to reach these goals men have to foster virtue. At the end of the day, virtue ethics is about making a good person who does the right thing spontaneously.

For Aristotle, virtue (*arete*) refers to excellence of some sort that can be of two kinds: moral (learned from habit) or intellectual (learned from teachers). Virtue ethics aims at becoming a good person, which means that one has to think about what is worth being considered as a virtue. In *Protagoras* and *Republic*, Plato offers four cardinal virtues: prudence, temperance or restraint, courage or fortitude, and justice. These virtues can be found, albeit with other names, in Aristotle's *Nicomachean Ethics*. According to Aristotle wisdom is an intellectual virtue, while prudence, justice, temperance and courage are moral virtues. Be that as it may, virtues are not given from outside, they have to be chosen by each individual.

Aristotle stresses that moral virtue 'is a state of character concerned with choice, and choice is deliberate desire, therefore both the reasoning must be true and the desire right, if the choice is to be good, and the later must pursue just what the former asserts.'[50] Thus it would be impossible to claim that obedience to orders is a virtue and to impose it upon soldiers as in the *Code du soldat*. Yet, looking more closely, one could say that obedience is about courage; courage to disobey if necessary, as well as courage to assume responsibilities resulting from disobedience.

It is pretty common within the French armed forces to assume that orders are morally just because they have been ordered by legitimate authorities who are implicitly supposed to have evaluated their morality. This transforms Socrates' phrase about determining whether 'the pious or holy is beloved by the gods because it is holy, or holy because it is beloved of the gods' into the following question: 'is what is morally good commanded by the Authority because it is morally good, or is it morally good because it is commanded by the Authority?'[51] This is a tough question which a soldier cannot answer.

Defining which virtues are worth developing requires deep thinking. It would certainly seem obvious, for instance, to say that courage is a central virtue for a soldier,[52] but it is not enough to act courageously because it is the 'right thing' to do; one must also have the proper dispositions in doing so.[53] Aristotle stresses that virtues are to be chosen for themselves or for the sake of happiness.[54] In so doing,

---

50 Aristotle, *Nicomachean Ethics*, translated by W. D. Ross (Kitchener, Ontario: Batoche Books Limited, 1999), VI: 2, p. 92.

51 Plato, *Euthyphro*. 10a.

52 Robinson, 'Integrity', p. 38.

53 Aristotle, *Nicomachean Ethics*,VI: 12, p. 103.

54 Ibid., I: 7, p. 10.

he insists on the role of wisdom as a virtue. According to Aristotle, 'it is thought to be the mark of a man of practical wisdom to be able to deliberate about what is good and expedient for himself, not in some particular respect, ... but about what sort of things are conducive to the good life in general.'[55] Deliberation is thus important, for one must not merely obey low cost moral rules. Practical wisdom requires understanding. Reflection implies a responsibility to develop one's moral sensitivity and ability, i.e. to keep on learning throughout one's life, since virtue ethics is not about guidance, and will not prescribe what to do in a particular situation. Virtues are corrective in that they help us to deal with difficult situations, to find an acceptable solution to moral dilemmas. Obedience to orders is a good example, and virtue ethics would help finding a golden mean, a 'mean between two vices' as Aristotle puts it. If one accepts that obedience is about courage, then courage is the mean between cowardice and foolhardiness, and obedience must be a mean between full submission (or passive obedience) and disobedience. This mean is nothing else than 'active obedience'.

Virtue ethics therefore dictates that soldiers are not expected to passively learn from repetition of orders and then act in accordance with this training. Rather, they are expected to develop a critical spirit, thanks to which they will learn from their own mistakes and develop good character. It is through practical wisdom (*phronesis*) and rational principle[56] that they will act in a virtuous way and then be moral, not through obeying orders without judging them. A lack of good moral training will lead both to soldiers being unable to form correct moral judgments and to them passively obeying orders that could lead to moral mistakes.

Conclusion: A Call for Nietzsche's "Free Spirit"

Ethics is a complex and demanding field of studies. There is something weird about pretending to do ethics while contenting oneself with a superficial approach to moral philosophy. Thus it seems hypocritical to issue moral codes devoid of serious moral underpinnings. It seems that in the French armed forces ethics is becoming merely a catchy tool for communication. Ethics speaks to everybody, but not everybody is able to speak about ethics.

Returning to the *Code du soldat*, as we have seen so far it is disputable whether it can qualify as a moral code. Although some will find the formalization of Kantian deontology in this kind of code, it is not enough. Deontology cannot and must not be used as a policy tool to constrain soldiers. It must not turn into low cost deontology. It must be explained, it needs to be taught in all its dimensions instead of merely being used to teach soldiers 'to obey even petty and foolish commands'[57] without questioning them. Emile Durkheim captured it insightfully when he wrote

55 Ibid., VI: 5, p. 94.

56 Aristotle, *Nicomachean Ethics*, pp. 27-8.

57 Walzer, *Just and Unjust Wars*, p. 311.

that 'to teach morality is neither to preach nor to indoctrinate; it is to explain.'[58] Kant never called for the submission of some to others. On the contrary, he insisted on the need for each individual to be able to decide autonomously through rational reasoning which rule can be universalized and give birth to a categorical imperative. More than that, he encouraged reflection, whereas applying rules which one does not understand has nothing to do with reflection. On the contrary it is the best road to sclerosis of the mind. One must avoid creating situations where a military man 'obeys instantly', without hesitation, without arguing.[59] Hesitating is sometimes useful for hesitation can be used as a momentum for reflection.

Moreover, deontology, let alone low cost deontology, is not a goal in itself. Even if France, as well as Europe, is generally Kantian, utilitarianism and virtue ethics must not be left on the side of the road.

Low cost deontology, by allowing soldiers to argue that they 'just obeyed orders' is a strong incentive for freeing them from their responsibilities. This must be avoided at all costs. Soldiers have to be held 'responsible for what they do'.[60] Contrary to Walzer's claim that the military man's 'goal is to perfect an instrument of obedience; the uses to which that instrument is put is beyond his responsibility',[61] soldiers have to feel responsible, and in order to understand and assume their responsibilities they must think about the consequences of their actions. Here utilitarianism is undoubtedly of great help. But like deontology it involves deep thought and the ability to reason. This means that soldiers need to learn to use all ethical notions with increasing sophistication in order to accurately evaluate their acts and behaviours.

It is clearly the responsibility of all soldiers, whatever their rank, to be sure that their acts will always be done according to the principle of utility, 'which approves or disapproves of every action whatsoever, according to the tendency it appears to have to augment or diminish the happiness of the party whose interest is in question.'[62] It is therefore obvious that, in this framework, selective conscientious objection can be seen as a duty for soldiers who are expected to think about the consequences of their acts and decisions.

When it comes to the moral obligation to obey orders, the concern 'is not so much about disobeying as about being able to assume one's responsibilities.'[63] The problem is that disobeying orders is too often seen as a mark of a lack of professionalism, if not a complete absence of it, whereas it is 'rather a high manifestation of it', since, as Martin Cook writes in a compelling way, 'the very essence of a profession … is that the professional does possess and exercise

58 Emile Durkheim, *Moral Education* (Mineola: Dover, 2002), p. 120.

59 Huntington, *The Soldier and the State*, p. 73.

60 Walzer, *Just and Unjust Wars*, p. 311.

61 Huntington, *The Soldier and the State*, p. 73.

62 Bentham, *An Introduction*, I :II, p. 14.

63 de la Maisonneuve, *Le métier de soldat*, p. 14. Author's translation.

independent judgment.'[64] This applies even to a soldier's participation to a particular war.

Soldiers need to learn to decide which virtues are worth developing. They need, for that purpose, to give sense to their actions. Then they have to practise. But one must be clear about that: it is not a matter of drill, i.e. the mere repetition of exercises aiming at making soldiers obey instantly. It is about the practice of virtues that have been chosen, not imposed from the outside. If one agrees that disobedience is a risky decision, and that courage is about making risky decisions, then one needs to develop courage. Individuality is at stake here. As Mill warned, people, in which we must include the military, 'may be progressive for a certain length of time, and then stop: when does it stop? When it ceases to possess individuality.'[65]

Following Arendt one must not mix up consent with obedience. The latter is for children, the former for adults.[66] Since soldiers are not children they are expected to obey actively, which means not to consent to or apply orders resulting from low cost deontology. 'There are ways of responding to an order short of obeying it',[67] and the moral soldier must seize any opportunity he has to do so if he considers the orders he has received to be morally disputable. General de la Maisonneuve writes that he learnt from General de Gaulle that 'beyond obedience and the worship of discipline … the soldier in charge of national independence is first and foremost a free man.'[68]

Military ethics needs to cut the apron strings of low cost deontology and favour other ways of thinking about ethics. Responsibility needs to be placed at the core of the profession of arms. It is necessary to offer military personnel room for intellectual freedom, and to open widely the debate over military ethics to explore new horizons. This is the only way to active obedience, to General de Gaulle's 'man of character'. Even more, it is the only path toward Nietzsche's 'free spirit'. Facing free spirit, arguments against selective conscientious objection are no longer valid. Soldiers must exercise their autonomy and express their free will even if it will sometimes clash with their superiors' orders.

For Nietzsche, human beings need to free their minds from institutions' constraints, 'from the crowd, the many, the majority',[69] through knowledge. So too did he call for the advent of the 'free spirit', 'who thinks differently from what, on the basis of his origin, environment, his class and profession, or on the basis of the dominant views of the age, would have been expected of him. He is

64 Martin L. Cook, *The Moral Warrior. Ethics and Service in the U.S. Military* (Albany: State University of New York Press, 2004), p. 63.

65 Mill, *On Liberty*, chapter 3, p. 66.

66 Arendt, 'Personal Responsibility', p. 46.

67 Walzer, *Just and Unjust Wars*, p. 314.

68 de la Maisonneuve, *Le métier de soldat*, p. 17. Author's translation.

69 Friedrich Nietzsche (1886), *Beyond Good and Evil*, translated by Helen Zimmern (Rockville: Serenity, 2008). 2:26, p. 30.

the exception.'[70] There is nothing more valuable than free spirits, for they help institutions to advance, even if sometimes they embarrass their hierarchy.

Deontology, utilitarianism and virtue ethics, all together, call for freedom of will and reflection. They all call for knowledge, as Nietzsche did. It is time to escape from low cost deontology. With this in mind, one can see that selective conscientious objection can be considered legitimate, if not a moral obligation, when one is asked to fight a war which hurts one's moral convictions. The argument which says that obedience imposes a duty on every soldier to agree to fight whenever the political power decides is no longer morally valid. Considering that obedience is not a moral obligation, soldiers, on behalf of their integrity and moral courage,[71] are thus permitted, and perhaps clearly required, to refuse to fight in particular conflicts with which they do not agree.

70 Nietzsche, *Human All Too Human*, 5: 225, p. 108.
71 Robinson, 'Integrity'.

Chapter 3

Selective Conscientious Objection: A Violation of the Social Contract

Melissa Bergeron

The right of conscientious objection is acknowledged by most of the nations still conscripting military personnel,[1] although how difficult it is successfully to exercise this right of conscience varies widely by country and circumstance.[2] This putative right is attributed to various sources: an extension of the right to life; a requirement for protection of the young; or freedom of thought and conscience. These need not be taken as mutually exclusive. I find the last of those justifications the most compelling, but my purpose in this chapter is not to evaluate any particular defence of selective conscientious objection. Instead I argue that what makes defences of universal conscientious objection[3] compelling is the implicit appeal to respect for human dignity, which is not an applicable defence in the case of selective conscientious objection, the objection to participation in a particular war.

Someone already serving as a combatant cannot coherently argue that she is being pressed to serve in a manner fundamentally contrary to her most developed and reflective moral identity, a claim clearly available to, say, the Quaker who objects

---

1 Or, among nations which are not actively drafting, but employ some registration mechanism intended to facilitate conscription should that become necessary, e.g. the United States of America's Selective Service System. See also note #2.

2 The *International Covenant of Civil and Political Rights* (1966) mentions 'conscientious objection/conscientious objectors' [Article 8(3)(c)(ii)] but seems to grant wide discretion to States to recognize (or not) this right. Not until Resolution 1987/46 by the Commission on Human Rights is the final wording cast in terms of freedom of thought, conscience, and religion, though it primarily 'appeals to states' to follow the counsel of the Resolution. Conspicuously absent is guidance regarding appropriate responses to the deliverances of conscience once military service is already underway. The appeal found in the 1987 Resolution is transformed into 'recognition' by the 1989 Resolution (59). And by 1992 the Commission published a Special Rapporteur on human rights and youth noting both that it is a fundamental right to refuse to kill and, that States should recognize the rights of individuals to be released from service on grounds of conscientious objection (UN Doc E/CN.4/Sub.2/1992/104-107). For an excellent discussion, see H. Takemura, *The International Human Right to Conscientious Objection to Military Service and Individual Duties to Disobey Manifestly Illegal Orders* (Heidelberg: Springer, 2009), chapter 2.

3 According to universal conscientious objection an objector takes issue with fighting in any war, not just to fighting in a particular war.

to being conscripted. The very act of entering into service in the first place requires that one's considered moral judgment allows for the use of violence in the service of the state, which is a willingness absent in the case of universal conscientious objection. *A fortiori*, the injury the state inflicts on the universal conscientious objector by forcing him to employ violence as a means to state ends cannot be true of the combatant who objects only to participation in a particular war.

In skeletal form, then, my argument is that it is the responsibility of the individual combatant to ensure that she is not subordinating her moral agency to an institution not suited or willing to justify that faith and allegiance. Once she does enter into that agreement, it is legally binding. I do not wish to deny that it might be morally incumbent upon a fighter to refuse to follow lawful orders she finds morally offensive. But that does not mean that it should be legal for her to do so. It is crucial to note that public refusal of this sort, as with civil disobedience, constitutes an important political mechanism facilitating change. Legalizing selective conscientious objection would preclude this sort of political act and constitute elimination of a crucial opportunity for correction. The fact is, and should remain, that she is bound by the law; the terms of her contract were perfectly clear at the outset, 'obey all lawful orders'. Under no reasonable construal is the law perfectly coextensive with morality under the best of circumstances, let alone in the context of war. Indeed it is challenging enough to ensure that law and morality are consistent.

This line of reasoning presumes that a combatant understands what is expected of her when she commits to service, takes the requisite oath, and undergoes the relevant training. She thereby becomes a dangerous person, as Michael Walzer describes it.[4] Furthermore, if she is not to be guilty of moral abdication in allowing herself to become, quite literally, a lethal instrument of the state, then she must have good reason to believe that her sovereign will not betray that trust, will not exploit her obedience, employing her in the service of unjust ends. I appreciate that this is a high standard, perhaps unique in its demands. But soldiering is unique among vocations, the only one, for instance, in which the foreseeable deaths of innocent people do not necessitate that one desist.[5] Thus a demanding standard is not just merited; in context, it is essential.

This argument turns on two additional assumptions. Following others in the contemporary liberal tradition,[6] I take consent to be the only legitimate basis for

4 M. Walzer, *Just and Unjust Wars: A Moral Argument with Historical Illustrations* (New York: Basic Books, 1977), p. 145. One is a combatant in virtue of 'his having been made into a dangerous person' and 'for that reason he finds himself endangered.'

5 The Just War Tradition allows for the possibility that the foreseeable killing of non-combatants might be permissible.

6 Very broadly construed to include the egalitarian pluralism of John Rawls through the contractualism of Thomas Scanlon to the more communitarian leaning Michael Sandel. While there are interesting and important differences among these theorists and their ilk, discussion of those differences is unnecessary for my purposes. A coarse appeal to social

a ruler's exercise of power. Mine, broadly, is a social contract theory. I grant that, regarding some nations, the social arrangement might seem a particularly anaemic form of self-governance, possibly involving regimes only just meeting the criteria for 'minimally just states'.[7] I believe that one might articulate a general notion of a self-determining people sufficient to make the use of social contract theory apt and serviceable.[8] A people must achieve a certain level of self-determination to count as 'a people'. And it is that power which makes a people ultimately responsible for the sovereign they have and (by extension) what is done in their name. In any case, I do not have in mind borderline cases. In connection with this, my final assumption is that a polity need not be perfect in order to be adequately responsive to (and representative of) the people's will. The citizenry of a minimally just state can make its will known and heeded if there is a mind to. When a people fail to act justly it is a failure of will, not of power. I will have more to say about this when I discuss the lack of sovereign accountability (*jus ad bellum* failures) which I believe gives rise to the perceived need for selective conscientious objection.

The Bifurcation of the Just War Tradition

The Just War Tradition assigns moral responsibility in war according to the role of the agent, rendering the sovereign responsible for waging war (*jus ad bellum*), and individual combatants responsible for how well that war is fought, morally speaking (*jus in bello*). This division of responsibility has the net effect of morally exempting individual combatants for the fact of their participation in war, irrespective of the justice of the war. Combatants are thus legally permitted to kill enemy combatants, even if the 'enemy' is, in fact, the defender. Jean-Jacques Rousseau, offering one of the more bluntly stated rationales for this division, insists that 'war is a relation, not between man and man, but between State and State, and individuals are enemies only accidentally, not as men, nor even as citizens, but as soldiers; not as members of their country, but as its defenders'.[9] Lawful combatants[10] on both sides of the conflict are equally blameless for their

---

contract theory is sufficient to make my case, and the usual difficulties associated with social contract theory, viz., the nature and so binding force of the contract, do not apply here, since the contract at issue in my argument is literal and explicit.

7 B. Orend, *The Morality of War* (Peterborough: Broadview Press, 2006), p. 35. Basically that both the populace and that state's neighbours view the sovereign as their legitimate ruler and his administration makes every reasonable effort to satisfy their material and political needs.

8 See Rawls' *Law of Peoples* where he entertains a variety of government forms along the spectrum of peoples, from a constitutional democracy to a non-liberal but 'decent' regime.

9 J. Rousseau, *The Social Contract & Discourses* (London: Everyman, 1973), p. 187.

10 Possibly a fuzzy category, though international law and the war convention seem basically satisfied with features of the combatant that prevent confusion with non-combatants, e.g., openly carrying arms, wearing uniforms or some clear designation of one's status.

participation in war, the Tradition insists, when they target only one another. They are, in Just War Tradition parlance, morally equal.[11]

This sense of moral equality emerges out of a rather specialized notion of 'innocent', which distinguishes combatants from non-combatants and makes lawful targets of any members of the first category.[12] The moral equality granted all combatants is not a judgment regarding moral worth but of blameworthiness. Those in the business of harm, as Michael Walzer casts it, are (by virtue of their training and willingness to fight) not innocent in this sense, even if individually they do nothing morally wrong. All combatants, by definition, are in the business of harm, and so are endangered in virtue of the danger they pose. The moral equality of combatants holds that a combatant's fighting is not her own doing and so morally cannot be counted against her, since:

> When soldiers fight freely, choosing one another as enemies and designing their own battles, their war is not a crime; when they fight without freedom, their war is not their crime. In both cases, military conduct is governed by rules; but in the first the rules rest on mutuality and consent, in the second on a shared servitude.[13]

Compliance with the rules is what falls to individual combatants and constitutes the sole basis for judging them. Individual combatants are not legally permitted to refuse participation in a war even if they came to believe it was an unjust war.

It strains credulity for one to claim that the soldier offered his services to fight only in just wars, so that the harm he suffers in being compelled to remain a combatant according to his contract is akin to the harm the universal conscientious objector suffers. *Ex hypothesi*, the combatant is not being forced into combatant status contrary to his most developed and reflective moral commitments, as with the universal conscientious objector. If this were true, then the most it would establish is that his decision to enter service in the first place was hasty and imprudent. This would be a culpable lapse.

Professional boxers, as a simple illustration, surely do not think that a pugilist, in virtue of entering into the ring, is a killer. Yet, the nature of the sport is such that the failure to recognize and reconcile this possibility constitutes culpable

---

11 The term used by Walzer is the 'moral equality of soldiers', which has since evolved into the more inclusive 'moral equality of combatants'.

12 By the other members of that category.

13 Walzer, *Just and Unjust Wars*, p. 37. Note that Walzer's notion of war is broader than Rousseau's. The latter would not grant that soldiers designing their own battles are warring, properly understood, since 'war is constituted by a relation between things, and not between persons; and as the state of war cannot arise out of simple personal relations, but only out of real relations, private war, or war of man with man, can exist neither in the state of nature, where there is not constant property, nor in the social state, where everything is under the authority of laws.' (Rousseau, *The Social Contract*) Still, both agree that the role of the combatant is what determines the scope of her moral responsibility.

negligence. This is a sport the express objective of which is to render one's opponent unconscious via repeated strikes to the head and torso. The potential of inflicting serious (possibly even fatal) injury is non-negligible. It would be irresponsible to ignore this possibility and determine in that light whether the danger is sufficient to undermine the overall morality of the endeavour. I gather that most judge that the risk is not great enough, that there is sufficiently robust consent from all parties, that there are ample safety measures and oversight, and so on, and therefore that it is a morally acceptable enterprise.

Analogously, warfighting is, by its very nature, a lethal endeavour. The force combatants employ does not merely have the potential to kill; it is the express purpose of the training received and the tactics employed to maximize the harms inflicted, swiftly and decisively defeating the enemy. And it is a virtual certainty that there will be collateral damage, if only because modern wars occur where non-combatants live.[14] One may hope that fighters wish to fight only just wars, but it is unlikely that they will routinely be in a position to know the status of a war at the time when they are called upon to prosecute it. It is naive to suppose that a combatant could be certain, or even relatively confident *in the moment*, that her skills are used only in the service of justice.

To cite but one example: given an order to fire on coordinates where the blast will also foreseeably kill non-combatants, a soldier working field artillery cannot know whether this is in fact a permissible course of action, satisfying the criteria of the doctrine of double effect. The military relies on a rigid hierarchical command structure precisely because it is impracticable to attempt to justify orders to the moral satisfaction of each link in the chain. If this example is a case of permissible collateral damage (surely a possibility), it would likely not be obvious to the individual soldier carrying it out. For this reason (and others) he is required to obey all orders unless he can be sure they are unlawful. The moral equality of combatants is a policy that, in part, fortifies the division between the decision maker and the person executing that decision, leaving each agent attending to that which he is best placed to manage.

While I think more must be said about the modern justification for the moral equality of combatants which (it seems to me) has nothing to do with invincible ignorance, incorrigible patriotism, or any other such grossly implausible pronouncements, the deeper problem, and the one to which I shall attend, lies in the utter lack of penalty, and so restraint, attaching to misdeeds of the sovereign in the modern world. As Thomas Hobbes sagely observed, 'Covenants, without the Sword, are but Words, and of no strength to secure a man at all.'[15] The contemporary

---

14 Some figures set the loss of non-combatant to combatant life at 10:1. Probably this was more a reflection of the destroyed infrastructure and the disease, famine, and post bellum chaos that made the death toll of innocents so high. Still, conservative figures have anywhere from a 3:1 to a 5:1 civilian to soldier ratio.

15 T. Hobbes, *Leviathan*, ed. Richard E. Flatman and David Johnston (New York: Norton, 1997), p. 93.

breakdown in Just War Tradition is at the *ad bellum* level, for nothing compels the sovereign to satisfy the criteria for a just war, nothing but the demands of his people. And the pressure exerted by a disgruntled electorate is the form such demands take, viz., a citizenry exercised over the clear injustice of the war their sovereign wages.

The cost of foot soldiers second-guessing civilian leadership is (in my judgment) both imprudent and, given certain social contract obligations, unjust, especially if this paves the way for even less accountability on the part of the sovereign, as I shall argue it would. To allow a combatant to leave the service according to selective conscientious objection would undermine the legitimate authority of the sovereign to determine which wars a nation wages and, crucially, would also undermine the responsibility attendant upon that authority. Moreover the mass exodus of the more astute moral reasoners certainly would not help matters in a war that surely would proceed, their absence notwithstanding.

More to the point, the very basis for demanding selective conscientious objection – that the sovereign's decision to wage unjust wars constitutes a harm which combatants (among others) should not have to endure – seems to neglect the obvious: it is the job of the sovereign to ensure that a nation wages only just wars. In the case of the United States, the sovereign (at least in this respect and by law) is predominantly Congress. There is clear, public record of who votes for what. Congress holds constitutional authority to raise and maintain a military force. Congress alone holds the legal authority to declare war and, more to the point, holds the purse strings. Importantly, each chamber is the product of popular vote,[16] making the connection with the electorate as clear and direct as it gets in a representative democracy. In short, the structure for holding accountable the sovereign is in place, functional, and adequate to the task.[17] The lack of control over the sovereign[18] lies in the will and moral fortitude of the people, the citizenry, not (specifically) the soldiery.[19] Increasing the pressure on combatants by legalizing selective conscientious objection is an unacceptable strategy for addressing this injustice. It is the apathy of the citizenry that is the cause of a democracy's waging unjust wars. If the reason for proposing selective conscientious objection in the first place is to protect combatants from the great wrong of an unjust war, then there can be no better way than to prevent (or, barring that, halt) the war. And if the

16 Actually, the 17th Amendment, which established direct election of senators, allows for the appointment of replacements in case the office is vacated mid-term; the specifics vary from state to state. I set aside this oddity as, ultimately, inconsequential regarding the general directness of representation.

17 Though it is worth noting that Congress has not elected to exercise its obligation of oversight and representation on behalf of the People since the Second World War. See Paul Yingling, 'How Congress Has Abdicated Its War Powers', *Armed Forces Journal*, February 2010.

18 Which, as an analytic matter, rather needs to amount to lack of self-control.

19 I use this term to cover all combatants.

war is unjust, then many more people beyond the objecting combatant are harmed: the enemy combatants they kill, the innocent bystanders caught in the fray, and the citizens whose safety, station, and good name is blemished are also harmed.

The social contract reasoning which best makes sense of contemporary states establishes, in general terms, the division of labour that we find in the Just War Tradition. Given this structure, the role of the combatant in the context of an unjust war is analogous to the role of the citizen in the context of an unjust law: each, ideally, will stand against injustice, and each will likely suffer the consequences for violating a law, even if it is an unjust one. The lack of *ad bellum* accountability cannot reasonably be expected to improve as a consequence of selective conscientious objection (or virtually any other alteration at the *in bello* level). If the moral concern is the waging of unjust wars, as it should be, then the sovereign inactions are the problem, not the binding obligations of the combatant. And if combatants are willing to serve their country by killing and possibly dying on the orders of a commander, it hardly seems excessively demanding or unreasonable to expect that those same soldiers muster the moral courage sufficient to endure the legal hardships attaching to noncompliance.

Social Contract Theory and Individual Responsibility

Social contract theory, at least in its contemporary form, presumes a division of labour similar in many respects to the bifurcation found in the Just War Tradition. The relationship supposed of citizenry and sovereign is contractual, a compact whereby the sovereign organizes and executes tasks not effectively or plausibly managed by the whole of the populace (at least not directly); the citizens, in essence, provide (through tax dollars, resource allocation, and personnel) the means for carrying out sovereign intent.

In the United States, as noted above, the will of the people rests most clearly and fundamentally with Congress, which (not incidentally) holds the most immediately significant powers to enact law, declare war, tax and spend, etc. It follows, by design, that Americans are their own sovereign (at least in the full course of things). We are constitutionally, and by extension morally, legislators unto ourselves. Only via a relatively healthy and responsible Congress can the actions of the sovereign truly be an expression of the people's will.

I belabour this point, because it is often suggested (or perhaps just supposed) that the citizenry, relative to the government, is best characterized as bearing rights (i.e., holding claims against the government). While this is true, it is also misleading. In addition to the rights citizens enjoy, there are responsibilities that secure those crucial rights. Voting stands out as a bland but serviceable illustration. To say that voting is a right picks out only one aspect of a citizen's role under the social contract. Cast this way, it does not serve to highlight the obligation I share with my fellow citizens to exercise oversight, to ensure that our collective will is carried out. Voting is a duty, in this way of thinking, as much as a right.

Similarly, it is true that I may protest an unjust law. But this seems to make light of an awesome responsibility. My own intuition is that where injustice is at issue the obligation comes to the fore; one *must* stand against injustice. We tend to think of citizens who court risk in lobbying for social change as being commendable for their courage. This emotional disposition is understandable, perhaps even admirable. But treating all cases of risk-courting as supererogatory obscures the obligations we share to be decent and responsible members of our community. When fellow citizens are being brutalized by an unjust system – our system – it is our shared and individual obligation to speak out. This is the inescapable consequence of life in a community, particularly a constitutional democracy.

Understandably, we are hesitant to demand of one another considerable personal sacrifice. But at some point, one must. Perhaps we cannot condemn our neighbours who did not participate in sit-ins in the segregated south, protest unjust wars, occupy Wall Street, and so forth, especially if they would have had to shirk other moral responsibilities or bear disproportionate burdens in order to take such risky stands. But this is surely a conservatism justified by our epistemic limitations and not the moral reality. Since I cannot know just how costly it would be for you to protest against some particular injustice, how secure your awareness of the injustice is, whether you are up to the task, and so forth, I am rightly less confident in my conviction that you should take a stand than in my conviction as to whether I should stand against it, and less confident still that you should be penalized or judged for your perceived failing. But that I cannot judge with confidence whether you discharged your moral obligation to the rest of us, given our shared political existence, is not the same as there not being such an obligation. Epistemic access is one thing, moral fact is another.

Even if we cannot say with precision just when a 'may' becomes a 'must', even if we are not always willing to punish or shame those who fail to risk their own wellbeing for a random person on the street, it does not follow that we can never make such judgments. Add training, experience, or a fiduciary obligation, and the uncertainty falls away.

If we sincerely wish to ensure that sovereign nations wage only just wars, this requires that the sovereign tasked with the responsibility to make precisely this determination actually is held accountable in some demonstrable and (one hopes) effective manner. And if one believes that the sovereign cannot be trusted to discharge that moral responsibility, then it is incumbent upon would-be combatants to refrain from volunteering altogether for service (best case) that they have no reason to believe will be just, let alone (worst case) have ample evidence will be unjust. A private citizen has the obligation to consider the confidence she might reasonably place in her sovereign's *ad bellum* behaviour; for once she enters service, her legal power to act on her individual moral sensibilities (in this matter) is greatly attenuated, and rightly so.

An individual should refuse to participate in anything that she judges immoral, anything that undermines her integrity as a moral agent; but it is also her obligation as a member of a polity that she obey the laws democratically ratified or suffer the penalty for noncompliance. This is not an inconsistent claim, though it might be a tragic state of affairs. If she cannot in good conscience obey a law, then she satisfies her obligation under the contract by making her disobedience a public stand against the injustice, by making her case to the rest of us. Any contract has the structure of a hypothetical, is 'if-then' in nature, such that if she will not serve as agreed, then she will be charged, in effect, with breach of contract. In this scenario, she is not merely privileging her own interests or even her good conscience over her obligations to the collective. Rather, she is challenging the community. She is, by her actions, demanding that the polity explicitly reject her concerns, choose the law in question over her moral sensibilities, and (perhaps the most crucial piece) punish her for that challenge, for her infraction. There surely can be no more explicit evidence that the community is, after all, secure in that law, human costs notwithstanding. She, by this act, forces the test that proves reflective community acceptance.

As John Rawls explains it, 'civil disobedience is a public act', intended to stimulate social discourse, for 'not only is it addressed to public principles, it is ... engaged in openly with fair notice; it is not covert or secretive.'[20] The central point is that, name notwithstanding, civil disobedience is a civically functional activity, a political act. Life in the state is, fundamentally, a collaborative enterprise, majority rule under agreed-upon principles, emerging out of a shared conception of justice. It is worth considering at length Rawls' articulation:

> By engaging in civil disobedience one intends, then, to address the sense of justice of the majority and to serve fair notice that in one's sincere and considered opinion the conditions of free cooperation are being violated. We are appealing to others to reconsider, to put themselves in our position, and to recognize that they cannot expect us to acquiesce indefinitely in the terms they impose upon us. ...
>
> Indeed, civil disobedience (and conscientious refusal as well) is one of the stabilizing devices of a constitutional system, although by definition an illegal one. Along with such things as free and regular elections and an independent judiciary empowered to interpret the constitution (not necessarily written), civil disobedience used with due restraint and sound judgment helps to maintain and strengthen just institutions. By resisting injustice within the limits of fidelity to law, it serves to inhibit departures from justice and to correct them when they occur.[21]

The assumptions giving rise to wide acceptance of universal conscientious objection are intuitive: to compel a person against his nature to violent behaviour is itself a sort of violence. To make, in essence, a pacifist[22] fight is to change who

20 J. Rawls, 'Definition & Justification of Civil Disobedience', in Hugo Bedau (ed.), *Civil Disobedience in Focus* (New York: Routledge, 1997), p. 106.

21 Rawls, 'Definition & Justification', p. 114.

22 See *Gillette v. United States* (and *Negre v. Larsen*), though for an interpretation possibly in tension see *United States v. Sisson.*

he is, not just what he does, and this violent change occurs no matter the justice of the enterprise. Selective conscientious objection, by contrast, would allow citizens already serving as combatants[23] to object, not to violence per se, but to the sovereign's decision to use state violence to a particular end, to use them to a particular end in a particular circumstance. This may be an injustice but, if so, it is a different one from that of denying universal conscientious objection. Denial of selective conscientious objection is not to compel to violence someone who is opposed to that very type of act, to change that person. This is a case of insisting that someone already amenable to violent means (in the service of the state and executed on command) employ these means in a particular scenario. It may well be that the justice of the objective will affect the person charged with pursuing that end. But it cannot plausibly be claimed that the very pursuit is damaging, that the means employed unproblematically in other settings have, themselves, become injurious. The justification appropriate for universal conscientious objection does not apply to selective conscientious objection.

Combatant Responsibility and the Division of Labour in the Just War Tradition

A policy allowing selective conscientious objection would undermine sovereign powers and accountability in a way that universal conscientious objection cannot. As a contingent fact, there are relatively few universal conscientious objectors, if history is a guide. And opting out of the whole enterprise of legitimately employing violence for state ends is not the same as hand-wringing over specific episodes of state force. Selective conscientious objection cannot but undermine the division of labour that characterizes the Just War Tradition, i.e., *ad bellum* and *in bello*. And to blur the division of labour also is to conflate the responsibility attaching to the relevant agents.

It is worth noting that the person fighting in what appears to be a just war cannot be sure that its status will not change, that human error or corruption higher up the chain will not make her an unwitting instrument of injustice. This is not an appeal to ignorance but a blunt reminder that one function of the moral equality of combatants is to secure precisely the sort of protection that makes possible this level of trust, which is essential to each doing her part. Combatants are always at risk in this way, succeeding or failing as a corporate body. It is an inescapable extension of the obedience necessary to managing such large scale, complex endeavours as war. As a practical matter, it is hard to see the principled basis we might employ for establishing limits on the time, place, and interval of combatant reassessment of sovereign decisions were we to grant selective conscientious objection. Would the combatant be able to object to a sovereign decision only in some temporal proximity to the decision, like a statute of limitations? What if she had no objection to the initial act, an invasion perhaps, but subsequently judges

23 I set aside, for the moment, conscripts and those who are yet to be inducted.

that the moral status has changed? After all, one of the *ad bellum* criteria is the probability of success, which might change at any moment. May she revisit the matter later? With what frequency? I think it is clear that this is untenable.

A citizen should not have entered into service in the first place if she cannot be relatively confident in the moral fortitude and constancy of the sovereign. And even if there was no reason to expect that the sovereign would behave unjustly, it is still possible that the sovereign might betray her. But she, in that case, is every bit as much a victim as the people she kills, provided she exercised due diligence in deciding to place that measure of trust in her sovereign. And if she comes to believe that her sovereign has betrayed that trust, then her choice is simple (and awful): either she fights in what she believes is an unjust war, making her morally complicit (but not legally culpable), or she refuses, making her prima facie legally culpable (but not morally complicit). But if she enters into service with no evidence as to the justice of the sovereign, then she is complicit in the wrongs she helps perpetuate. While enlistment is not wholesale abdication of moral autonomy, it is sufficiently modulating of one's liberty to demand very careful consideration regarding the prospects for due care on the part of leadership, weighed against the possibility of exploitation and mismanagement.

This division of moral labour works, if it does, not merely because it allows each to do her part: It forces it. The pressure from within, the harm a sovereign does to his people (as a people) and the growing dissatisfaction of the citizenry in the throes of that harm, is an essential part of the health and functioning of a well-ordered polity. Selective conscientious objection reduces this important pressure. As with civil disobedience, it is the unjust suffering of our own, as much as, or even more than, our own good conscience that prompts change and correction. But we must be made to confront the suffering that follows from our unjust actions. And for this, the suffering must obtain and visibly so. If the young combatant sits in the brig rather than serve in an unjust war, then the pressure is on, and his action is not inconsequential, is not a matter of removing himself entirely from the power structure which makes possible our warring.

Prisoners of Conscience

The brute fact is that in fielding soldiers we are willing to expend their lives in the service of state ends. And our combatants are willing to kill the enemy on the same assumption. I propose no more than this, indeed considerably less. More wars than not are unjust. If a combatant fights in a war she suspects is unjust, then she must reconcile her compliance (and subsequent actions) and her conscience. If she refuses to fight, paying the price for that refusal, then she serves her country still, just not with her life or her moral innocence. I confess, I do not see how we can overlook the value of the objector's conscience and its power to right our course, a power that does not come at the cost of innocent lives, hers or those she might kill. To eliminate her challenge – time served making public statements of refusal, appearing in court, maybe even being sentenced to time in the brig, for taking the moral high ground – is to reduce the pressure on us to do the right thing.

We need fighters, people who are obedient and do what they are told, immediately and without reservation, for sometimes wars are just. But we also need people whose lives reflect back to us, intimately and at home, what we do when we wage wars unnecessarily. Selective conscientious objection precludes this. It takes our more refined moral agents out of both categories: they are not mirrors, speaking truth to power, for they avoid penalty in their refusal to fight. As a result, the injustice they might be quite right about is no more highlighted for the rest of us than it would have been had they just fought; their insight is impotent to effect change. And, we end up dumbing-down, morally speaking, the service personnel, the lower ranks of today who would be the commanders of tomorrow, thereby further diminishing the moral core of our fighting forces.

To enter into military service is to subordinate oneself to the collective will. It is inescapably a great moral risk for any number of reasons. A firmly bifurcated Just War Tradition, buttressed by the moral equality of combatants, is part of a system that locks people into especially rigid roles, roles that guide, constrain, and (perhaps perversely) comfort. And it also serves to establish some limits on what might be done to combatants for their participation, given the necessary limits placed on their access and power. Selective conscientious objection unlocks those roles, opens the combatant to what can only become an implicit requirement (however informal) to consider the wars in which they fight. If you are free to leave and do not, then the rest of us are free to judge you for your choice, for your failure to leave. But in their capacity as soldiers, combatants are not fundamentally individual citizens, entirely free to function as autonomous persons. They are agents of the republic. While their humanity is not dormant, their individual personalities and perspectives in large part are.

This is not meant to neutralize moral autonomy altogether, so much as to bracket or modulate individual expression of it. In the presence of great 'howling injustice',[24] we rightly expect the individual to find herself at odds with the group, slipped from the corporate body. As Walzer notes, though the enemy 'alienates himself from me when he tries to kill me, and from our common humanity ... the alienation is temporary, the humanity imminent.'[25] And I suppose that, in a sense, we are each mildly alienated from the humanity within us, at least in its guise of robust moral individuality, when we subordinate ourselves to the general will. But it is only one aspect of that humanity that goes dormant, and then only within bounds.[26] The latent humanity, the stable moral core, is a safety mechanism, of sorts, that springs when things go awry.

24 A resonant turn of phrase used by Wertheimer: R. Wertheimer, 'Reconnoitering Combatant Moral Equality', *Journal of Military Ethics*, 6/1 (2007), pp. 60-74.

25 Walzer, *Just and Unjust Wars*, p. 142.

26 It is worth noting that however many high ranking officers (e.g., JCS Gen. Pace, 2/17/2006) insist that combatants are enjoined to refuse 'an order that is either illegal or

Refusal to serve in clear cases of *ad bellum* injustice is not the same as simply opting out. A crucial feature that distinguishes the two is that refusal-with-penalty forces the citizenry to confront their injustice, assuming that the objecting combatant is right. Even if the combatant is wrong, it will compel a polity to reflect and clarify its commitment to the war. This can hardly be a bad thing. And if we do value this as a safety mechanism, then we cannot afford to cut loose exactly those moral agents with sufficiently rich sensibilities to serve in this capacity. The great responsibility incurred in taking the oath, which one hopes was taken neither lightly nor naively, remains even in the wake of sovereign malpractice – perhaps more so then. Combatants, those disposed to obey the sovereign fully, constitute the best evidence that we are in the wrong (should they object to our warring) and our best hope of righting that wrong. The last thing we should be willing to do as a people is to eliminate that crucial safeguard to our collective moral wellbeing.

We sometime suggest, in lamenting the moral insufficiency of holding combatants to *in bello* restrictions alone, that the morality of war is a unitary matter, susceptible (ultimately) to a single moral judgment, and that the Just War Tradition perverts justice by parting what should be left whole. Personally, I think that in forcing *ad bellum* obligations onto combatants we might well facilitate abandonment of both standards, though I grant that this consequentialist worry (even if accurate) does not engage the objection from injustice, which I take the selective conscientious objection to be primarily concerned with. If we restrict attention to the matter of justice, then we should recognize that the more pernicious system malfunction is the command of a cavalier or immoral sovereign and the attitudes and beliefs fostered under such command. If an Ivy League attorney working for the president can transform torture into legal orders in a few speciously reasoned pages; if rules of engagement permit the targeting of a man known to have just a camera (not a weapon); if terror, degradation, and abuse constitute the standard which the Commander-in-Chief sets, unchecked by the sovereign despite the clear unconstitutionality of such behaviour; then the lesson to learn is not that individual combatants should have greater latitude (and so increased moral obligation) regarding the orders of civilian leadership.

Rather, the worry is that this sort of leadership is no leadership at all. The individual combatant is responsible for adhering to certain procedural standards. She is not granted licence to dispense with elements of that system, to refuse to follow orders, according to the end she hopes (or anticipates) it will achieve. Even if a combatant can see that better results might be attained by violating the relevant rules, this alone (even if true) does not justify dispensing with procedure. The sovereign, by contrast, is charged with adjudicating among competing moral values and aspirations. While the sovereign does not have unfettered latitude, in

immoral', this is not what the law says. 'Moral' is entirely absent. Even the notion of 'lawful' reasonably expected of the average foot-soldier is greatly limited, usually understood by the courts as requiring one to refuse to commit 'grave breaches of law', which is a considerably narrower swath of unlawfulness.

contrast to the individual combatant it must consider the likely consequences of selecting a particular course of action. Where the combatant's discretion is greatly constrained, the sovereign's is not. And with this additional power comes expanded responsibility. Indeed, one might argue that the authority vested in the sovereign is for the very purpose of expanding the field of available options. Where the fighter is enjoined to follow procedure come what may, the sovereign is expected to activate processes according to the ends they reasonably can be expected to secure. One would never surmise this difference from knowledge of sovereign behaviour over the past decade.

What I have argued for is simply the status quo, although I have attempted to articulate the social contract underpinning of the Just War Tradition bifurcation that (if I am right) precludes selective conscientious objection. I do not pretend for a moment that this is an ideal outcome, that we should not be every bit as exercised by the clear exploitation of fighters, employed in unjust wars, as in fact some of us are. Rather, my position is that, from a practical standpoint, it is untenable to institute a policy of selective conscientious objection. Individual combatants cannot reasonably be expected to function under this additional burden; and it is a burden. There is, so far as I can tell, no principled basis in a selective conscientious objection world for determining just when a combatant's appraisal would rightly come to a close.

In addition to the practical challenges associated with a policy of selective conscientious objection, it is unclear how it might contribute to a just system. A system which allows individual combatants to opt out of particular wars is a system which thereby makes them implicitly accountable for their participation in war, not just their specific actions therein. An already difficult division of labour would leave neither of the two sorts of agent wholly and solely responsible for his role, for there would no longer be clear and distinct roles. There is overwhelming evidence that the conduct of the USA over the past decade of war has been manifestly unjust, including a rich variety of violations of our obligations under the Geneva Conventions, and sundry other instances and episodes of manifest malfeasance on the part of the American sovereign.[27] As a people, we have left wholly unchecked the antics of our sovereign, which has allowed the bald seizing of power by the executive branch, and sent the military unjustly into the spaces and behaviours now causing such worry. Demanding, on top of this, even greater responsibility of the individual fighters, neither empowered nor situated to discharge effectively this new dimension of obligation, is nothing short of perverse (and a recipe for disaster).

The obvious solution seems to be to fix the part that is broken. In the United States, for instance, Congress needs to spend perhaps a little less time congratulating itself for its leadership and start exercising some of the oversight it is sworn to execute. This would include, primarily, reining in an executive branch that seems to have lost sight of the scope of its constitutional authority as well as its status as one among three distinct branches of government.

---

27 See *Hamdan v. Rumsfeld.*

Chapter 4
Who Guards the Guards? The Importance of Civilian Control of the Military

David Fisher

In the early 1990s I spent much time, together with NATO colleagues, lecturing and organizing workshops in the capitals of former Warsaw Pact countries on civilian control of the military. For the former Warsaw Pact generals and other officers the democratic accountability of the military was a strange and alien concept. They had been rather accustomed to telling the politicians what it was they could and could not do, rather than being told. It is a feature of totalitarian regimes that, even where they are not overt military dictatorships, the military are, nonetheless, generally accorded a privileged position since it is on their support that the dictator relies for his continued exercise of power.

It was, therefore, hardly surprising that the initial reaction of the former Warsaw Pact officers was often suspicion and distrust of the notion that the military should be obedient to civilian politicians. It was, nonetheless, considered vital that this lesson should be learnt since the principle of the subordination of the military to the civilian political authority was regarded as fundamental to the establishment of democracy in those states. This same lesson is perhaps being learnt in countries emerging from the Arab Spring. In Pakistan the military and civilian authorities are engaged in their perennial struggle for power.

The importance of the principle has, however, come under challenge in recent years from those philosophers, such as Jeff McMahan and David Rodin, who have argued that the traditional view of the moral equality of combatants is mistaken. On that traditional view combatants on either side have an equal right to use force and an equal obligation to use force justly, complying with the constraints of *jus in bello*. But that view, it is argued, is mistaken. Combatants on the just side are entitled to use force against those fighting an unjust war because the latter are morally responsible for the injustice of the war. There is, however, no such moral license for unjust combatants to use force against those fighting a just war. It is only combatants on the just side who are entitled to use force. There is no such entitlement for warriors on the unjust side. Service people have, therefore, a moral obligation to refuse to fight in an unjust war. In order to enable them to exercise their moral duty they should be granted a right of selective conscientious objection. They should be able not just to decline to fight in war if they believe war itself to be immoral, as is already conceded by the right of general conscientious objection. But they should also have the right to pick and choose which wars they fight, refusing to fight in any they consider to be immoral.

A right of selective conscientious objection would enable citizens in a state where there is conscription to decline military service not just because of a general conscientious objection to war but also because of objections to a particular war. In Australia following a public debate provoked initially by the unpopularity of the Vietnam War and subsequently reignited following the First Gulf War, the Defence Legislation Amendment Act was introduced in 1992 (with amendments in 1993) which granted to conscripts a right to seek exemption from participation in a particular military conflict on the grounds of seriously held moral convictions.[1] The right has, however, never been exercised since the Australian Government abolished national service in 1973, moving to an all-volunteer forces, for whom no such right was granted. Any increase in the exemptions allowed to conscription raises questions over the appropriate balance to be struck between the rights of an individual and those of the community to defend itself. It may also raise concerns over fairness where the burden of military service is distributed compulsorily amongst parts or the whole of the male citizen body (or in the case of Israel, male and female).

But in countries such as the UK and US where there are all volunteer forces the choice whether or not to join the military is already voluntary. A right of selective conscientious objection would, however, enable individual service people, whether conscripted or volunteer, to decline to participate in a particular war of which they morally disapproved, while still continuing to serve in the forces. Should such a right be conceded to service people in a democratic society, particularly in all-volunteer forces of the kind increasingly the norm in Western states?

Granting such a license to the military to disobey the orders of the civil authority would represent a major change in the relationship between the military and civil authority, hitherto judged of importance to democracy. Powerful arguments would, therefore, be needed to justify such a change. So are there such arguments?

It is argued that holding soldiers fighting for an unjust cause morally responsible for the war's injustice more accurately reflects our moral thinking than the traditional just war doctrine of the moral equality of combatants. It is also suggested that reinterpreting just war thinking in this way would have decided moral advantages. Attributing moral responsibility in this way, coupled with the concession of a right for an individual service person to decline to fight in a war he or she considers unjust, will encourage service people to think much more carefully about the wars in which they fight. This will discourage them from fighting in unjust wars and, in turn, will make it more difficult for politicians to wage unjust wars. A substantial moral gain would thus be achieved.

Let us consider each stage of this argument and, first, why these philosophers have sought to challenge the moral equality principle. According to the traditional

1 Lieutenant Colonel Ian Wing, 'Selective Conscientious Objection and the Australian Defence Force', *Australian Defence Force Journal*, 137 (1999). http://www.adfjournal.adc.edu.au/UserFiles/issues/137 1999 20Jul_Aug.pdf.

account a soldier is entitled to use defensive force against other combatants because of the threat of harm that they pose. Force is legitimized to stop the harm occurring.[2] But posing a threat of harm does not, in McMahan's view, make one liable to attack. For if that were true a policeman about to shoot a gunman threatening the lives of citizens would render himself liable to defensive attack by the gunman. The gunman, however threatened he may feel, has no right to take defensive action against the policeman. 'It is therefore,' McMahan claims, 'false that by posing a threat to another, that necessarily makes one liable to defensive action.'[3] What makes one so liable is 'moral responsibility for an objectively unjustified threat of harm.'[4] 'The act of posing the threat has no relevance apart from the contribution it makes to the person's moral responsibility for the threat.'[5]

So applying this to war, what makes soldiers and, indeed, some civilians liable to attack is that they are morally 'responsible for a wrongful war or for the wrongful acts of which it is composed.'[6] The combatants fighting a just war have no such moral responsibility and so are not liable to attack. But combatants fighting an unjust war are morally responsible for its injustice and are hence liable to attack. The doctrine of the moral equality of combatants is, therefore, false.

A key step in this argument is the supposition that in non-military contexts we always require an attribution of moral responsibility to justify the use of defensive force. According to McMahan the police can use force against the gunman because he is morally responsible for the harm being caused and so has rendered himself liable to defensive attack. But the gunman has no such right to use force against the police who bear no such responsibility. That sounds plausible. But let us examine more closely the respective roles played by moral responsibility and the threat of harm in determining against whom defensive force may be used.

Let us suppose that a gunman is on a hotel balcony spraying bullets down into a crowd gathered in the square below, which includes young children. There are many casualties. The senior police officer in command of the operation judges that the only way to stop this mayhem is to shoot the gunman. A police marksman takes aim and is about to fire when news is brought that the gunman is an innocent dupe who is being coerced to act by an evil third party. The evil third

---

2 David Fisher, *Morality and War: Can War be Just in the Twenty-first Century?* (Oxford and New York: Oxford University Press, 2011) pp. 83-4.

3 Jeff McMahan, *Killing in War* (Oxford: Oxford University Press, 2011), p. 14. The arguments for and against abandoning the doctrine are explored in David Rodin and Henry Shue (eds.), *Just and Unjust Warriors – The Moral and Legal Status of Soldiers* (Oxford: Oxford University Press, 2010). The classic statement of the case for moral equality is in Michael Walzer, *Just and Unjust Wars: A Moral Argument with Historical Illustrations* (New York: Basic, 1977), pp. 34-41.

4 McMahan, *Killing in War*, p. 35.

5 McMahan, *Killing in War*, p. 208.

6 *Killing in War,* p. 204. This is McMahan's shorthand version of the criterion. A fuller version is: 'The criterion of liability to attack in war is moral responsibility for a war that may permissibly be prevented or corrected by means of war, or by an act of war', p. 204.

party has hypnotized the gunman and, under hypnosis, instructed him to open fire on whoever enters the square. The evil hypnotist is, however, now distant from the scene of action and peacefully watching television in his flat. The hypnotist is morally responsible for the threatened harm. So would the police be justified in rushing in and shooting him? The answer is that they would be justified in doing so only if the evil hypnotist were somehow still exercising influence or control over the gunman's activities. If – as we have assumed – he is not currently exercising such influence, they would not be justified in shooting him since such action would not stop the gunman's murderous activities. It is the removal of a threat of harm that justifies the action, not the moral responsibility of the person for the harm.[7] Moral responsibility is thus not a sufficient condition for the use of force.

But nor is moral responsibility a necessary condition for the defensive use of force. This can be seen by considering another variant of the gunman example. There is on this occasion no evil third party but, as the police marksman is about to fire, the gunman's wife appears, distraught, on the scene. She explains to the officer in command of the operation that her husband is suffering from paranoid schizophrenia and in his delusion believes that inner voices are instructing him to mow down the evil forces with whom he mistakenly believes he is confronted. His mental illness could be accepted by a court as an excusing condition which, while not justifying his action, would remove or at least diminish his responsibility for it. He would lack the necessary *mens rea* to be held responsible. So having learnt these new facts should the police marksman be instructed to stand down? The answer would appear to be yes if the justification for the police's use of defensive force was the moral responsibility of the gunman. We demur, however, from reaching such a conclusion since it would allow the gunman to kill many more people. We think rather that the police would still be justified in using force in order to prevent the harm the gunman is causing or threatening to cause other citizens. So the police marksman should open fire.

The person morally responsible for a harm and the person posing a threat of harm will normally be one and the same. So it will usually be true that it is legitimate to use force against the person morally responsible for the harm. But where, as in these examples, the concepts diverge it is the person posing a threat of harm against whom the defensive action can legitimately be taken since that is the way to remove the threatened harm.

Is the moral responsibility of the agent then of no significance? On the contrary, it is crucial to our moral appraisal of the action. It is also crucial to determining

---

7 McMahan argues that to protect his life a Mayor would be justified in killing an evil Sheriff who is morally responsible for a threat to his life rather than the gun-toting hired farm hand who is actually posing the threat. But in order to make killing the Sheriff appear justifiable McMahan has to hypothesize that the Sheriff's death will remove the threat to the Major posed by the farm hand who 'will feel himself released from his assignment.' (McMahan, *Killing in War,* p. 206). It is the removal of the threat that justifies the killing. Moral responsibility would not suffice to justify this.

whether and how the gunman should be punished. But the policeman in the midst of a dangerous operation, still less a soldier in the heat of battle, are neither best placed to make such fine judgements of moral responsibility. Neither the police nor soldiers are appropriate organs of punishment. We accordingly leave it to the courts subsequently to determine questions of moral or legal responsibility, which are crucial to the apportionment of punishment.

Force can only justifiably be used to punish someone if he or she is morally responsible. Moral responsibility is a necessary condition to justify the punitive use of force. The punishment would otherwise be unfair. In early just war thinking the use of military force was justified as a form of extra-territorial punishment. But that view always faced serious difficulties, not the least being that military force is typically used against ordinary service people, rather than the political and military leaders who, as we shall explore further shortly, might more legitimately be held morally responsible for the harm. War as punishment would be grossly unfair. There are thus good reasons for rejecting a punitive justification for the use of military force. But there are other justifiable uses of force than the punitive.

The defensive or protective use of force can be morally justified without necessarily requiring the prior attribution of moral responsibility for the harm. Force can justifiably be used against someone threatening harm simply in order to stop that harm occurring. The police are not licensed to exact punishment. Nor are they required to determine who bears moral responsibility. What the police are licensed to do is to protect citizens from threatened harm so that 'normal life can continue.'[8]

But if force can be justifiably used to stop harm occurring, why is the gunman not entitled to use force to defend himself against the police marksman? The answer is that we regard the gunman's killing spree as a harmful activity that needs to be stopped, whereas the policeman's action is not so regarded. This judgement in turn derives from the fact that we cede to a government of a state a monopoly on the use of physical force to maintain order. This is a key step in the civilizing process of moving from a violent world of private vendettas to an ordered and law-governed society. It is the process described so memorably by Aeschylus in the *Oresteia* as the avenging Furies are eventually brought under the control of the Athenian judicial system. The chosen organs to exercise that monopoly are the police for internal order and the armed forces for external order. A logical corollary of ceding such a monopoly to the police is that a private citizen who is posing a threat of harm to other citizens has no right to use physical force to defend himself against a policeman using proportionate force to stop that harm occurring. The state concedes to citizens a right of self-defence in certain circumstances. But concession of such a right against a policeman carrying out his duty to protect citizens would be self-

8 National security is 'a state of confidence on the part of the citizen that normal life can continue, despite the dangers to individuals, families and businesses from man-made threats or impersonal hazards'. David Omand, *Securing the State* (London: Hurst and Company, 2010), p. 17.

defeating and undermine that monopoly. We authorize the police to use on our behalf necessary and proportionate force, including the use of deadly firearms, against someone threatening harm in order to protect citizens from harm; an authority that is not conceded to ordinary citizens. We require, moreover, that the general license is carefully applied by police officers in individual cases to ensure that any force used is proportionate and necessary to achieve the objective of stopping harm.

Let us now consider McMahan's application of his argument to the military context. McMahan concludes that most combatants fighting an unjust war are morally responsible for the wrongful war or for the wrongful acts of which it is composed. He concedes that a few may have such strong excuses (irresistible duress or justified ignorance) as to evade responsibility. But most unjust combatants do not have such strong excuses.[9] They may have been subjected to some coercive pressures but not to unbearable duress, still less duress that is literally irresistible. Nor can they be excused through ignorance since, even if they are ignorant, their ignorance may still be culpable. Soldiers typically do not give enough thought to the wars in which they are engaged but should be encouraged to do so. So most soldiers can legitimately be held morally responsible for the injustice of the wars in which they are engaged. Moreover, if they are held to be responsible, they will be less likely to engage in war.

Can ordinary service people legitimately be held responsible in this way? We hold people morally responsible for actions over which they have some influence or control; and to the occurrence of which they have, in some sense, given their consent. Control and consent are the two key preconditions of the attribution of moral responsibility. The generals who advised Hitler may have had such influence over events and consented to their occurrence. They can, therefore, legitimately be held morally responsible for the injustice of the war both as a whole and its component acts. But that is not true of the ordinary soldiers who had no influence or control over the decision to go to war. It strains the concept of moral responsibility beyond legitimate usage to hold most combatants on the unjust side responsible for the injustice of the war. Nor can they be held responsible for its component acts in so far as these reflect the injustice of the war rather than the individual choices of combatants over how the war is conducted. To hold ordinary service people morally responsible for the injustice of the war would seem neither appropriate nor fair. It was for this reason that the Nuremberg tribunal rejected the argument that *all* military actions undertaken by the Germans were war crimes because they were carried out in pursuit of an aggressive war.[10] That judgement was well founded, however much we may debate whether the tribunal may – for reasons of political expediency or practicality – have been too selective in those it held to account.

9 His detailed anatomy of soldiers' excuses is in McMahan, *Killing in War*, chapter 3 'Excuses.'

10 *Trials of War Criminals before the Nuremberg Military Tribunals* (Washington: US Government Printing Office, 1950-51), xi. pp. 488-9. http://www.loc.gov/rr/frd/Military_Law/NTs_war-criminals.html.

Barcelona in 1936 but did little to enhance the military efficiency of the Republican forces. Indeed, the ill discipline of the Republican forces contributed to the eventual success of Franco's forces and so to the triumph of fascism in Spain.

Obedience is also fundamental to the relations between the military and the civil authority within a democracy. The obedience of the military to the lawful orders of the civil authority is critical to the effective functioning of democracy. Such obedience can never be taken for granted. As Samuel Finer observed in his sombre survey of the role of the military in politics:

> There is a common assumption, an unreflecting belief, that it is 'natural' for the armed forces to obey the civil power ... But instead of asking why the military engage in politics, we ought surely ask why they ever do otherwise. For at first sight the political advantages of the military *vis a vis* other civilian groupings are overwhelming. The military possess vastly superior organisation. And they possess arms.[16]

Finer was writing in 1962 when within Western Europe, only recently recovered from Nazi tyranny, there were still military dictatorships in Spain and Portugal. There had been a successful military *coup* in France as recently as 1958 and an unsuccessful military *putsch* against the Paris Government by the French generals in Algeria in1961. There has since been an extension and stabilization of democracy in Europe, not least thanks to the workings of the European Union. But in Eastern Europe and Russia democracy is still a recent and fragile institution, while elsewhere in the world military takeovers continue with depressing regularity. Since 2012 a fierce struggle has been taking place in Middle Eastern countries to wrest control from the military and so establish democracy following the revolutions of the Arab Spring. Subordination of the military to political authority is a hard won principle that, as such experience elsewhere in the world amply illustrates, is fragile and needs constantly to be defended, if we are to avoid military tyranny.

Obedience is thus important. But even in a democracy there may be occasions where the requirements of justice may require disobedience. Let us consider first the requirements of *jus ad bellum* by examining two contrasting examples. Most of Hitler's generals, despite their post war protestations to the contrary, supported Hitler's wars of aggression and were deeply implicated in the resulting atrocities. As Ian Kershaw has recently noted: 'Military leaders and rank-and-file soldiers alike were well aware that they had been responsible for or implicated in countless atrocities in the east –torched villages, mass executions of partisans, shootings of tens of thousands of Jews.' [17] The generals also continued waging the war from

16 S.E. Finer, *The Man on Horseback: The Role of the Military in Politics* (1962; London: Pinter Publishers, 1988), p. 4.

17 Ian Kershaw, *The End – Hitler's Germany 1944-45* (London: Allen Lane, 2011), p. 394.

the summer of 1944 in a doomed struggle that, particularly on the Western front, served no conceivable purpose but merely massively prolonged the suffering of all involved. When asked why they acted as they did the generals' answer was that they were obeying Hitler's orders. For most of them it was simply unthinkable to break their oath of loyalty and obedience to the Fuhrer. As Kershaw again observes: 'Most generals took their oath of allegiance to Hitler extremely seriously ... the traditional military imperatives of order and obedience were distorted in the Third Reich to an extreme readiness to yield to the commands of the Fuhrer, however irrational.'[18]

In the tyranny of the Third Reich there was no democratic option available to express dissent, while obedience to Hitler's orders resulted in the infliction of massive suffering. We, therefore, condemn the blind loyalty and unthinking obedience that Donitz, Kesselring and their fellow officers displayed to Hitler right until his suicide on 30 April1945. The generals should have opposed Hitler's decision to embark on a war of aggression; they should have refused to be involved in the murder of Jews; they should have declined to prolong a war they had already lost. Where a war is, in Vitoria's phrase, so 'patently unjust'[19] those with influence or control should have done whatever they could to prevent it, including, if all else failed, disobeying orders.

Contrast this with a recent case from the 2003 Iraq war. Should a British service person who believed the Iraq war was unjust have disobeyed the order to go to war? The decision to undertake military operations was taken by a democratically elected Government and approved by Parliament. The order was, therefore, lawful under the British constitution. The British Chief of Defence Staff, setting an important precedent for the future, had sought and received assurances from the Attorney General that the operation was also in accordance with international law, however much that judgement might subsequently have been disputed by other lawyers. The participation of an individual serviceman in military operations would not of itself have led to morally grave consequences. While, there were grounds for doubting the justice of the war, the issues were complex and open to dispute. There were also other options open to any officer in an all-volunteer force who felt compelled by his conscience not to fight. It would, in particular, have been open to him be to resign his commission at the earliest permitted exit point and leave the service.

Such a course could, for example, have been followed by Flight Lieutenant Kendall-Smith who objected to UK/US operations in Iraq. He voiced his objections in July 2005 by which time the coalition operations to maintain law and order had, unlike the initial invasion, been authorized by the United Nations under Security Council Resolution 1483, passed on 22 May 2003. Kendall Smith was advised that he could lawfully leave the service in 12 months' time. He was not, however,

18 Kershaw, *The End*, p. 395.

19 Vitoria, *On the Law of War*, 2.2, sect.22 in Anthony Pagden and Jeremy Lawrance (eds), *Vitoria: Political Writings* (Cambridge: Cambridge University Press, 1991), p. 307.

prepared to wait and chose instead to refuse to undergo training for operations in Iraq. He was court-martialled for failing to obey lawful orders and sentenced to 8 months in prison. If he had chosen instead to resign, this would have been to exercise, as an individual, a right to resign from military service, accepting whatever financial and other penalties might thereby be incurred.

In a democracy, with all-volunteer forces, resignation would seem in such circumstance a more appropriate course of action than conceding to ordinary service people a right to disobey a lawful order to go to war. Giving the military a right to pick and choose which wars they fight would risk undermining the fundamental democratic principle of the subordination of the military to civil authority. It would also risk damaging the cohesion, morale and efficiency of military units. The rights of individuals might be enhanced but this would be at the expense of the common good. We could never be sure when a unit was ordered into action how many, if any, would go. Such a reduction in military efficiency might appear acceptable, indeed welcome, if all wars are assumed to be unjust. But such a pacifist presumption has never been the view of the just war tradition, which has always acknowledged that war may sometimes be morally permissible and may, on occasion, be a moral duty. Impeding our ability to fight just wars, whether to defend the state from external aggression or to protect innocents from slaughter, would be detrimental to the common good and not represent a moral gain. This was a lesson that the ill-disciplined anarchists in Spain learnt to their cost from their doomed struggle against fascism.

With the recent shift from wars of necessity to wars of choice, it is, nonetheless, vitally important that our political and military leaders and the military at all levels are encouraged to think critically about the morality of the wars in which they are involved, as well as the morality of how they are conducted. Before war is declared there should be vigorous debate both in private and public about the morality of the war, a debate in which the generals and other military officers who advise Ministers should play a full role. Just war counsels should ring out in the Cabinet war rooms, as well as shaping public and Parliamentary debates. We should insist that all those with responsibility for a decision to embark on military action should give full regard to the justice of what is proposed. But expecting ordinary service people through the exercise of a right to selective conscientious objection to bear the weight of the moral decision to go to war is to place the responsibility in the wrong place with those who have little or no influence or control over the decisions.

On this question Vitoria, as so often, strikes the right balance. He insists that those who are privy to the king's counsels and 'who ... attend the public or royal council are duty bound to examine the cause of just war. If such men can by examining the causes of hostility with their advice and authority avert a war which is perhaps unjust, then they are obliged to do so.'[20] On the other hand, unless the

20 Vitoria, *On the Law of War,* 2.2, sect.24 in Pagden and Lawrance, *Vitoria: Political Writings*, p. 308.

war is 'patently unjust', ordinary citizens and service people 'may lawfully go to war trusting the judgement of superiors ... (for) they cannot prevent war even if they consider it unjust.'[21]

Similarly in a modern democracy the prime moral responsibility rests with those who have control or influence over the decision to go to war, particularly the political leaders and their civilian and military advisers. All those with influence over the decision to go to war, including the senior military, are 'duty bound to examine the causes of just war.' But in a democratic state once the decision to go to war is taken by the elected Government and authorized by Parliament, the presumption should be that the military will then obey the civil authority, and that lawful orders conveyed down the military chain of command will be obeyed.

Let us now consider the requirements of *jus in bello.* We need to ensure that our service people behave justly in the conduct of wars, not least since inhumane conduct can undermine the humanitarian objectives of a campaign. Military obedience should be reflective and not invoked for orders that are manifestly illegal or immoral which our servicemen should have the moral courage to challenge. Ordinary combatants cannot be held responsible for the justice or injustice of the cause for which the war is fought. But they can be held responsible for how the war is conducted over which they have influence or control. For an ordinary soldier, as Michael Walzer notes, 'when he violates the rules of war, superior orders are no defence. The atrocities he commits are his own; the war is not.'[22] The requirement to disobey unjust orders is thus of particular importance in the conduct of the war, for which service people, from the highest to the lowest level, may bear the full responsibility.

Let us consider some recent examples. In the autumn of 2003 in the aftermath of the coalition operations to overthrow Saddam's regime there were a number of cases of abuse by US and UK soldiers of Iraqi civilian detainees. One of those detained was Baha Mousa, a hotel receptionist, who was savagely beaten to death by British soldiers while held in a Temporary Detention Facility in Basra. Eight other detainees were also physically abused. The soldiers who carried out the abuse, as well as those who condoned it, should not have behaved as they did. They should have disobeyed any orders to carry out such abuse. Sir William Gage, who chaired the public inquiry into these events, expressed 'concern about the lack of discipline and lack of moral courage to report abuse within 1 QLR (Queen's Lancashire Regiment).'[23] He criticized the Colonel in charge of the regiment and even the padre who 'must have seen the shocking condition of the detainees, and the deteriorating state of the TDF (Temporary Detention Facility). He ought to have intervened immediately, or reported it up the chain of command but, in fact,

---

21 Ibid., 2.2., Sect 25, *Vitoria: Political Writings,* p. 308.

22 Walzer, *Just and Unjust Wars*, p. 39.

23 *The Report of the Baha Mousa Inquiry*, The Rt. Hon Sir William Gage (Chairman) (London: The Stationery Office, 8 September 2011), Volume III, Summary of Findings, paragraph 203.

it seems he did not have the courage to do either.'[24] A key recommendation of the Gage report is that we need to devise better ways to teach moral courage to the military.[25]

Moral courage is the virtue that a service person needs to challenge unacceptable behaviour and to disobey orders to carry out immoral or illegal actions. It is a difficult virtue to practise, particularly in an institution, such as the armed services, that rightly also prizes the virtues of obedience and loyalty. The exercise of moral courage may at times appear to conflict with that of such other virtues.

The difficult challenge required to exercise moral courage is illustrated by the agonizing of the US soldier, Adam Winfield, before he agreed to give evidence against his Staff Sergeant, Calvin Gibbs. From his appointment in November 2009, Gibbs had led a 'Kill Team' of which Winfield was a member and that had murdered innocent civilians in Afghanistan. In a letter to his father after the death of one such innocent, Winfield wrote:

> Pretty much the whole platoon knows about it. It's OK with all of them pretty much. Except me. I want to do something about it. The only problem is I don't feel safe telling anyone. The guy who did it is the golden boy in the company who can never do anything wrong and it's my word against theirs.[26]

Misplaced loyalty to his comrades and fear of the consequences of speaking out inhibited Winfield from doing what he knew was the right thing. Such misplaced loyalty had similarly led to the 'wall of silence' criticized by the judge in the Baha Mousa court martial from the officers and soldiers, all of whom had declined to give evidence against their comrades involved in the abuse of Iraqi detainees.[27]

For service people, operating in a culture where importance is accorded to obeying orders and being loyal to their comrades, it may present a particularly difficult challenge to recognize the occasion when an order should be disobeyed, and to act accordingly, despite the many countervailing pressures. This will require not just the exercise of moral courage but also of the virtue of practical wisdom – 'a habit of sound judgment about practical situations.'[28] If such abuses are to be prevented in the future higher priority needs to be accorded within the

---

24 *Report of the Baha Mousa Inquiry*, paragraph 112.

25 *Report of the Baha Mousa Inquiry*, Vol III, Part XVII, Recommendations, Recommendation 58.

26 Chris McGreal, '"Kill Team" US platoon commander guilty of Afghan murders,' http://www.guardian.co.uk/world/2011/nov/11/kill-team-calvin-gibbs-convicted, 11 November, 2011.

27 Aitken Report, *An Investigation into Cases of Deliberate Abuse and Unlawful Killing in Iraq in 2003-4* (London: UK Ministry of Defence, 25 Jan. 2008), p. 24.

28 Peter Geach, *The Virtues* (Cambridge: Cambridge University Press, 1977) p. 160. The importance of the virtues in military life is explored further in Fisher, *Morality and War,* chapter 6, 'Virtues'.

military to the virtue of moral courage. This will necessitate not only the better training recommended by the Gage Report but also a major culture change within the armed services.

Conclusion

Obedience is crucial to the efficacy of a military unit, while the obedience of the military to the civil authority is a fundamental principle of democracy. But disobedience may sometimes be morally required. In considering when disobedience may be justified we need to distinguish between the requirements of *jus ad bellum* and *jus in bello*. We should insist that our political and military leaders are well versed in just war thinking and that before military action is initiated there should be vigorous debate both within the private counsels of Government and publicly in Parliament and elsewhere to assess the justice of what is proposed. The military, at all levels, should be encouraged to think critically about the justice of the wars in which they are engaged. A particular onus should, however, be placed on those military leaders who advise Ministers and so share responsibility for the decision to go to war to ensure that war is only embarked on where it is just and the criteria of *jus ad bellum* are met. But once a democratic decision is taken by the civil authority the expectation should be that the military will obey. Disobedience would only be countenanced exceptionally where a war is 'patently unjust'.

Turning to the conduct of military operations, responsibility rests on all involved from the most senior officers to the most junior personnel to ensure their conduct complies with the requirements of *jus in bello*. Obedience is important for the successful conduct of military operations. But where obeying an order would have morally grave consequences disobedience may be not only morally justified but even morally required. Discerning the occasions when such disobedience may – exceptionally – be required will need careful thought and the exercise of the virtue of practical wisdom. Challenging unacceptable behaviour in the face of many pressures to stay silent also – critically – requires the exercise of moral courage, a virtue the importance of which needs to be recognized and accorded higher priority within military training and culture.

We need to improve the way we select, teach and train not only our political leaders and their senior civilian and military advisers but also our military at all levels to ensure that all those involved with war behave justly. But a legal right to selective conscientious objection is not the right way forward. It would weaken military efficiency and impede our ability to fight just wars. It would also risk undermining the democratic accountability of the military. As we taught the former Warsaw Pact officers in the 1990s, in a democratic state the military should, as a rule, be obedient to the civil authority. We put that fundamental principle of democracy at risk at our peril.

Chapter 5
An Empirical Defense of Combatant Moral Equality

Michael Skerker

This chapter attempts to reframe the debate about the moral equality of combatants in order to move it away from non-empirical arguments. Rather than focusing on service members' epistemic limitations, shared political servitude, or questions relating to a written or unwritten social contracts between the individual and the political community he or she serves, I will instead argue that the question of moral equality should instead turn on an empirical question of whether modern militaries can efficiently function while permitting service members to opt out of particular operations for reasons of conscience. I will argue that, as long as combatants cannot claim a right to selective conscientious objection, all lawful combatants fighting on behalf of a state *are* morally equal. Under these circumstances, the state institution separates service members' individual moral faculties from certain actions that they carry out. In a way, it removes those actions from an individually focused normative framework and replaces it with a corporate or collective perspective. At the same time however, service personnel also gain certain qualities in equal measure due to the nature of the institution they serve.

Framing the concept of combatant moral equality as dependent on an empirical question is a different approach than employed by nearly all the theorists who have addressed the topic of late, including those who defend the concept.[1] Rather than accepting the individual-centered approach which has dominated recent analysis, I will instead focus on the peculiar nature of the military institution, suggesting that the nature of service in this institution cannot simply be dismissed as morally irrelevant. As I will show below, a reframing of the issue in this way also has

1 For example, see: Michael Gross, *Moral Dilemmas of Modern War* (Cambridge: Cambridge University Press, 2010); chapters by various authors in Henry Shue and David Rodin (eds), *Just and Unjust Warriors* (Oxford: Oxford University Press, 2008), pp. 112-30; Lionel McPherson, 'Innocence and Responsibility in War', *Canadian Journal of Philosophy*, 34/4 (2004), pp. 485-506; David Rodin, *War and Self-Defense* (Oxford: Clarendon Press, 2002); Gerhard Øverland, 'Killing Soldiers', *Ethics & International Affairs*, 20/4 (2006), pp. 455-75; David Mapel, 'Innocent Attackers and Rights of Self Defense', *Ethics & International Affairs*, 18/1 (2004), pp. 81-6, and; Jeff McMahan, 'Innocence, Self-Defense, and Killing in War', *The Journal of Political Philosophy*, 2/3 (1994), pp. 193-221.

the benefit of moving past the controversial conceptual foundations of combatant equality frequently mentioned in recent literature.

The consensus for roughly the last 400 years in the Western Just War Tradition has been that all combatants obeying the laws and customs of war are moral equals. The meaning of this term has to be delineated. The causes for which they fight are not necessarily equally just and the leaders they obey are not necessarily moral equals. The combatants themselves may be unequal in terms of their courage, patience, honesty, temperance, and the like. There is moral inequality between combatants who commit war crimes, such as rape or perfidy, and those who refrain from such actions. The moral equality of combatants employs a restricted sense of morality regarding privileges and liabilities. During wartime, combatants on either side are equally permitted to kill their enemies at any time except following their surrender or incapacitation. They are also equally morally and legally liable to be attacked at any time. This is to say, neither side's service members—not the Germans or Poles, Israelis or Syrians, Americans or Iraqis—wrong the other by wielding lethal violence. Whether combatants fight for an unjust or just cause, the traditional argument is that they are neither consistently in the epistemic position to judge nor endowed with the professional authority to make determinations about the justice of the war. Since they are obliged by voluntary oaths of service or bonds of conscription to obey the orders of their superiors, the culpability for the broad corporate action they execute is transferred to the political leaders who have the knowledge and authority to make monumental foreign policy decisions.

However, in the last decade, a new consensus *against* combatant moral equality has arguably emerged in Anglo-American philosophical circles. Members of this consensus (called 'purists' by Yitzhak Benbaji) argue that combatants should not enjoy moral impunity and/or legal immunity for fighting in unjust wars.[2] They assert that combatants are responsible, to varying degrees, for judging the justice of wars in which they are called to serve, and for refusing to serve in unjust wars. The purists' conclusions differ from those of traditional just war theorists on this point because, unlike traditionalists, they do not assign warfighters special privileges on account of a corporate identity. Instead, they use the same individual-centered terms of moral analysis used in the analytic literature on self-defense to address questions of combatant culpability.

My argument against this approach is developed in detail in a forthcoming book where I posit that the average combatant has nothing like the knowledge, power, or intentionality of actors in self-defense scenarios.[3] Military actions are

2 See Yitzhak Benbaji, 'The Responsibility of Soldiers and the Ethics of Killing in War', *The Philosophical Quarterly* 57/229 (2007), pp. 558-72, and Yitzhak Benbaji, 'A Defense of the Traditional War Convention', *Ethics*, 118/3 (2008), pp. 464-95.

3 Michael Skerker, *Gunsoul: The Moral Status of Combatants* (University of Chicago Press, forthcoming 2014).

irreducibly corporate actions[4] and so cannot be scaled down for the purpose of moral consideration to miniature versions of themselves in the way, say, a group of senators stabbing Caesar can be scaled down to one senator stabbing Caesar. We have no problem assigning equal blame for the assassination to all conspirators, when we can scale down collective action that is the aggregate of largely identical individual actions. Individual-centered moral analyses, which assume that the actor has sufficient power to plan, commit, and complete an action, has complete awareness of the direct effects of his action, and a personal intention to commit the action, therefore finds sufficient material to analyze within the person of each senator. The group murdered Caesar; each senator is guilty of murdering Caesar.

By contrast, the morally interesting aspects of violent military action are lost when we look at a given service member's contributory action. This is because of the typical service member's fractional responsibility for corporate military action, his blinkered epistemic position on that ultimate complex action, and because he is compelled to adopt the military's corporate intention to achieve a particular objective. If an airstrike by an FA-18 is an example of a morally interesting military action, consider the thousands of people—defense contractors, politicians, intelligence analysts, targeteers, weaponeers, maintenance crew, fuelers, carrier staff, pilots—who causally contribute to a single airstrike with their vastly different types of contributory actions and different degrees of knowledge of the resultant action. Then compare the sufficiency of the action, knowledge, and intention of a Roman senator to accomplish the murder of Caesar with the sufficiency of the action, knowledge, and intention of say, the aircraft carrier's helmsman, to conduct the airstrike. Purists wrongly describe irreducibly corporate military actions when they effectively treat them as aggregates of identical individual civilian actions, such as the attacks of the Roman senators. This mis-description is an internal problem for purists who assume that culpability for unjust threats is necessary for liability to defensive violence.

Surprisingly, many purists do not explain how the injustice of a war inculpates service members for unjust actions; they simply assume that contributors to unjust collective actions must be committing unjust actions themselves. My critique about mis-description of corporate action is not necessarily a problem for the purists who do attempt to explain this connection between collective and contributing actions by articulating a notion of responsibility akin to strict liability in torts: an agent capable of responsibility (i.e. of sound mind) is liable for even indirect effects of his actions, regardless of his culpability for those effects. So a service member, who, acting under orders, marginally contributes to a war effort he non-culpably, but mistakenly believes is just, is liable to defensive violence. This follows, because he knew enlistment entailed a risk of being deployed to unjust wars. Very briefly, this inculpating standard is morally dissatisfying because

4 'Corporate action' has a technical meaning in the literature on collective responsibility: the action of a long-enduring organization with a fixed culture, executed as a result of a collective decision-making process such as a board of trustees' vote.

it imbues random events with the moral power to create liability for defensive violence. Strict liability in torts is not a direct expression of common morality in law as is the case with laws against murder or theft. Strict liability serves a practical function of ensuring self-regulation of dangerous but essential private activities like pharmaceutical manufacture, which the state can ill-afford to closely monitor. Since the pharmaceutical manufacturer is legally liable for tainted products even if someone completely unrelated to the manufacturer poisons a bottle of pills while it is en route to the store, the manufacturer is incentivized to take maximum precautions in the manufacture, shipping, and storage of the drugs. While it is unfair to the company to be penalized for a random instance of tampering it could not have reasonably foreseen or plausibly stopped, the law reduces the number of tampering cases overall by forcing all companies to exercise maximum—even excessive—levels of caution. While strict liability is close to the moral principle that one should use maximum caution when engaging in risky activities, we can see it is a tool of public policy rather than a moral principle because the scope of events for which the standard deems a person liable exceeds the scope of events foreseeably linked to the person's action. To ascribe moral responsibility for these events to the agent whose actions were factually among the causal roots for them is to effectively charge him—and in the military context, subject him to violence—with responsibility for random events.[5] Thus this method of inculpation is not a plausible one to deem service members liable to defensive violence by virtue of participating in unjust wars.[6]

If we can put to one side arguments *against* combatant equality (which perhaps the reader will grant, if just for the sake of argument), we still lack a *positive* argument in favor of combatant equality. This section will therefore present such an argument. Moving past a focus on individual moral qualities, we will now take into account the peculiar nature of military service which is traditionally considered relevant to Just War considerations. The nature of military service is morally relevant, if states or state-like organizations are considered necessary for

---

5 Jeff McMahan allows that this form of inculpation is subject to luck and therefore makes a slight, perhaps morally trivial difference between the person posing the unjust threat and the defender. However, that little difference can make all the difference in forced choice situations where one of two non-culpable parties has to suffer harm. See 'The Basis of Moral Liability to Defensive Killing', *Philosophical Issues* 15 (2005), pp. 386-404, 394. I think Seth Lazar is right that forced choice scenarios do not obtain in war, 'Responsibility, Risk, and Killing in Self-Defense', *Ethics*, 119 (July 2009), pp. 699-728.

6 Whitley Kaufman makes a similar argument in 'Torture and the 'Distributive Justice' Theory of Self Defense: An Assessment', *Ethics & International Affairs*, 22/1 (2008), p. 100. Also see Lazar for an excellent discussion of this method of inculpation. Lazar seeks to show the method's internal contradictions rather than criticize it for falling short of a standard account of accountability. He criticizes Kaufman for neglecting McMahan's earlier rejection of culpability's relevance to liability and failure to present an alternative account of liability. In my forthcoming book, I will defend the relevance of culpability to assigning responsibility for corporate action against McMahan's earlier rejection.

humans to thrive over time. All or most states need militaries;[7] and militaries are institutions with particular constitutive characteristics. I contend that there is a need to reframe the question of combatant equality in empirical terms on the basis of these considerations. Given their potential moral role in society, we have to ask if modern militaries can function if they extend to service personnel a right to refuse lawful orders, regardless of whether they concern accepting a particular military operational specialization (MOS—one's military job), deploying to a particular theater, conducting a particular operation within a deployment, or engaging in a particular tactical action within an operation. We could similarly ask whether the medical or legal profession could function, if we recognize some putative right or restriction on the part of doctors and lawyers.

One might object that selective conscientious objection is only about refusing orders to deploy to a war, but a right to refuse involvement in particular military actions would have to be broader than a right to refuse deployment. This follows since a more restricted right would not address Navy and Marine personnel on ordinary sea tours who are suddenly re-routed to participate in hostilities, or other service personnel based overseas who are ordered to conduct sorties from those points.[8] Furthermore, a controversial military operation potentially worthy of selective conscientious objection might consist of only a single mission lasting a few hours, such as the US raid in Pakistan that killed Osama bin Laden or the Israeli airstrike on the Syrian nuclear reactor. Effectively, a real right to selective conscientious objection would have to entail a right for pilots to refuse to participate in airstrikes, for helmsmen to refuse to adjust ships' headings, drivers to refuse to operate their vehicles, and so on.

Professional or professionally led militaries are legitimate state institutions if there are any just causes for wars necessitating organized actions more robust than *levee en masse.* If *standing* armies are necessary to create sufficient deterrence and fight just wars among modern states of a given size, then military service in standing armies in such states must in principle be permissible. A decision to enlist or accept conscription into a standing army is then a generically good one in that the enlistee or conscript is taking a potentially dangerous, service-oriented, modestly-paying job necessary for the long term good of his or her society. This generic good would be potentially outweighed, if the particular state in question has a long, sustained history of using its military exclusively for unjust wars and/or

7 It is possible for some small states to not need military protection because of their lack of strategic importance and/or lack of valuable resources and/or its enjoyment of safe environs guaranteed by a benign local hegemon.

8 Further, one can conceive of conscience-driven refusals to carry out specific tactical orders like US Army Capt. Lawrence Rockwood's refusal to desist from an investigation into prison conditions in Haiti in1994. Stephen Wrage, 'Captain Lawrence Rockwood in Haiti', *Journal of Military Ethics*, 1/1 (2002), pp. 45-52. Further, all lawful orders are equally obligatory, so there is no legal distinction between orders to deploy and orders to perform some tactical action with less significant consequences.

internal repression;[9] the government does not also successfully ennoble these actions with ubiquitous propaganda;[10] and the state is not surrounded by others with rapacious governments. In this sort of context, a citizen has grounds to see that military service to his state is not a generic good and that not even the potential defensive use of the military would justify enlistment, because the present state of affairs would be improved if the government fell to an aggressive but non-rapacious neighbor. A citizen is subjectively justified in enlisting if he has been thoroughly inculcated with government propaganda such that he non-culpably believes that his state is a basically just one entitled to defend itself. One who has objectively good grounds or objectively bad grounds (e.g. pervasive propaganda) for fearing worse governance if the current government falls is objectively or subjectively justified in enlisting, respectively.

The moral weight of enlisting during wartime is somewhat oversold, since the war may end before the enlistee is deployed or may shift from a just to an unjust war or vice versa. The strongest argument for enlisting even in the face of doubts about a war's justice involve the epistemic limits of a single citizen and the possible asymmetry between policymakers' and an ordinary citizen's knowledge of the strategic situation.[11] Nonetheless, it does seem correct to say that one should refrain from enlisting if one is convinced that one's country is fighting an unjust war.[12] This point derives from the broader claim that one should not do things one considers morally wrong, firstly, because they are wrong, and secondly, (even if one is mistaken about the moral value of the action) because acting against one's conscience violates one's moral integrity. In this case, an anti-war conviction would mean that the weight of the evidence for the war's injustice overcomes two cautions about the fallibility of judgment otherwise aiding pro-enlistment arguments. First, the convinced war opponent feels that the evidence for the war's injustice outweighs the potential qualifier that any one citizen's judgment about foreign policy is fallible. Second, the evidence also outweighs the potential counter-argument that a citizen should defer to his current political leaders even in the face of personal doubts since those leaders can be expected to be in a superior epistemic position. (A deeper level of skepticism about all government authority—and not just the current regime—would seem to preclude

9 David Mapel, 'Coerced Moral Agents? Individual Responsibility for Military Service', *Journal of Political Philosophy*, 6/2 (1998), p. 184; Jeff McMahan, *Killing in War* (Oxford: Clarendon Press, 2009), p. 72; Gross, *Moral Dilemmas*, p. 48.

10 Whether or not the government is popularly loathed would be a good external indication of successful propaganda.

11 Francisco de Vitoria, Vitoria, 'On the Law of War', in Gregory M. Reichberg, Henrik Syse, and Endre Begby, *The Ethics of War: Classic and Contemporary* Readings (Oxford: Blackwell, 2006), Q1.2; Walzer, *Just and Unjust Wars*, p. 39.

12 Christopher Kutz, 'The Difference Uniforms Make: Collective Violence in Criminal Law and War', *Philosophy and Public Affairs*, 33/2 (2005), p. 173.

enlistment unless one could be guaranteed that one would only be deployed to the war that motivated the decision to enlist.)[13]

The dilemma of serving personnel about to be deployed to a war they consider unjust is more fraught and philosophically complex than that of civilians contemplating enlistment since serving military personnel have taken oaths of service to the state. Volunteers also have a promise-based obligation to the state in return for the state's training investment and opportunity cost (for having not trained others who might not wish to leave the service prematurely). They have bonds of loyalty to, and responsibility for, the personnel they have trained alongside. Finally, if selective conscientious objection is untenable, they have a professional duty to follow legal orders, a duty which gets moral weight because of the service their profession renders society. This sort of professional duty to obey professional norms is akin to a doctor or lawyer's professional duty to maintain patient/client confidentiality or a police officer's duty to observe the norms of due process. These norms are primarily rule-consequentialist, oriented toward the efficient delivery of the professions' specialized skills and capabilities.

Service personnel must be permitted to take on the peculiar traits of military professionals, if standing armies are necessary and military service in them is therefore generically permissible (at least). In short, they need to be able to serve *as* soldiers, not as civilians in battle dress. This is why the question of combatants' individual culpability for judging the justice of wars and for refusing deployment in the case of unjust wars turns on an empirical question about the nature of modern militaries. It would be appropriate to use whatever individual-centered moral analysis one normally uses to judge individual actions and decisions, if one could serve in the military and still exercise a civilian's level of autonomy in judging a war's justice and deciding to give or withhold support. In this case the service member would have the same levels of knowledge, efficacy, and intentionality as a civilian and so would be as culpable as she would be for deciding to support an unjust war. Service members would be no different than attorneys who are free to assess the moral character of a potential client and refuse to represent him without losing their place in the firm. But this kind of analysis is not appropriate if permitting selective conscientious objection is practically untenable for modern militaries. If serving in the military and reserving the right to selectively refuse deployment (or MOS assignment or any lawful order) are incompatible, then a different mode of analysis tailored to the profession is required. This is, because service personnel *qua* their status as professionals cannot consistently have

13 A principled reason for enlistment by the person who rejects his current regime's authority would be in the reasonable hope that future wars his country might fight would be just. This enlistee might even want to aid in the current war for fear that a loss would embolden enemies and endanger his country. It seems hard to weigh this claim against a contrary one that argues a loss would be in the country's interest (the better to learn that unjust wars are imprudent). Given the small effect one's enlistment or non-enlistment would likely have, it seems reasonable here to defer to the concern about personal integrity.

all the knowledge, efficacy, and intentionality necessary for individual moral responsibility in executing lawful orders.

Combatant moral equality follows if military service is a generic good and selective conscientious objection is practically untenable.[14] In this case, service personnel in different states take on identical professional roles and with them the same corporate privileges and liabilities associated with those roles. If we imagine the different political systems, legal regimes, and wider cultures of a pair of antagonistic states represented by different kinds of architecture—say clapboard wood Colonials in one and brick row houses in the other—the militaries in both states would be represented by the same style building, say a squat stone ziggurat, strangely incongruous with its native surroundings and bearing more in common with its opposite number. It remains to argue that these privileges and liabilities entail killing and being killed, but such an argument would be related to an empirical question regarding how wars between states must be fought. Service members are morally justified in doing what is necessary to achieve national security goals like defense, deterrence, peace-keeping, and international criminal deterrence (for example, against piracy).[15] It is the answer to the question 'what is necessary' that will determine whether their national security purposes justify lethal violence. It is hard to imagine modern states abiding by the results of a non-lethal competition of the sort once practiced by some Native American tribes, but it is not a matter of conceptual contradiction. At any rate, all we are after here is whether combatants have equal privileges and liabilities, whatever they happen to be.

Why should the armed forces of many different states be conceptualized as the same kind of privilege-bestowing institutions, insulated from the variations in political systems and polices of the governments they serve? All conventional combatants in basically just states (and even unjust states with pervasive and effective propaganda) are in the same situation in that they sign up before they are sent to war or are drafted into an organization that depends on obedience to function. They do not know in advance about the justice of the wars in which they will fight, or about the possibility that an apparently just war they signed up to fight turns out to be unjust, or which transforms from just to unjust as circumstances and political goals shift.[16] Again, if selective conscientious objection is untenable, once in the military, service members from such states are equally bound by institutional norms and laws to deploy and fight, regardless of their assessment of the war's justice. These institutional norms get moral weight from the necessary role militaries play in basically just states. *Qua* their professional status, the motivation for the combat actions of service members on both just and unjust

14 McMahan, *Killing in War*, p. 72.

15 This justification serves as a basis for *in bello* restrictions. Murdering civilians, causing more suffering than is necessary for incapacitation, mutilating the dead, etc., does not directly contribute to these national security goals.

16 Ward Thomas, 'Unjust War and the Catholic Soldier', *Journal of Religious Ethics*, 35/3 (2007), pp. 509-25.

sides is the same—they are following orders traceable ultimately to a political entity. While they may have different personal motivations, prejudices, levels of commitment behind their actions, these are all collateral to their professional duty to obey lawful orders, which is to say, to use the military's corporate intention as a frame for forming intentions for their specific actions.[17] Military actions as such are caused by military orders and instructions.

Furthermore, all combatants are equally subject to their chain of commands' orders, strategy, tactics, equipage, training, MOS assignment. That is, other people's decisions which ultimately put them in the position to perform or contribute to the performance of morally significant violent actions. In war, they are equally part of a collective violence-threatening and violence-wielding enterprise that both diffuses their responsibility for, and blinkers their knowledge of, violent actions in the battlespace. This setting links their discrete activities with those of others to make them more threatening and dangerous than a civilian performing a similar action in isolation, such as firing a rifle. All of these qualities derive from the nature of military service and are independent of the justice of wars in which a military might participate.

At the same time, these qualities displace or efface those qualities necessary for personal moral culpability for various actions, which naturally vary from person to person. The service member is where he is, knows what he knows, intends what he intends, is able to do what he is able to do, and finally, contributes to the complex action he furthers, because of the institution he serves rather than his personal qualities.

There would be one group of service personnel exempted from this argument. Non-conscripted personnel in fundamentally unjust states with ineffective propaganda programs cannot claim to be doing a generically good thing by knowingly sustaining their brutal regimes, even if the regimes are threatened by (non-rapacious) enemies.[18] These combatants are not morally equal on an abstract level with service personnel from other types of states, but are closer to the status of gang members. I will beg the question of whether this abstract difference should make any difference in terms of their treatment in combat or in detention following capture.[19]

The remaining question is then whether selective conscientious objection is untenable? Although the need to reframe the moral equality debate as an empirical

17 See Seumas Miller, *Social Action: A Teleological Account* (Cambridge: Cambridge University Press, 2001) for a technical discussion of this sort of intention formation.

18 Perhaps they can be excused (at most) if the only alternative to military service is penury or if they remain in 'voluntary' service because of government coercion.

19 See Judith Lichtenberg, 'How to Judge Soldiers Whose Cause is Unjust', in Shue and Rodin, *Just and Unjust Warriors*, pp. 124, 129; Kutz, 'The Difference Uniforms Make', p. 175; David Rodin, 'The Moral Inequality of Soldiers', in Shue and Rodin, *Just and Unjust Warriors*, pp. 51, 56-60, and; Jeff McMahan, 'The Morality of War and the Law of War', in Shue and Rodin, *Just and Unjust Warriors*, p. 36.

question arises from the above argument, and moves the debate away from inherently contentious conceptual questions, the empirical question is not easily answered. As some point out, theorists are forced to speculate about the tenability of selective conscientious objection since apart from in Germany, modern militaries have never extended this putative right to volunteer service personnel.[20] If my argument has merit though, it will perhaps encourage other theorists to do further research in this vein and consider ways of modeling an answer to this empirical question.

To this point, I have found more data suggesting, in my mind at least, the untenability of selective conscientious objection. Arguing in favor of selective conscientious objection's tenability, some thinkers highlight two cases where relatively small groups of Israeli service personnel refused service in a particular war or particular type of operation without grave effects on the Israeli Defense Force. Of the nearly half million US service members involved in the Gulf War, official figures report, only 131 service members were discharged on conscientious grounds—presumably some of these were also really objecting to the Gulf War in particular instead of experiencing a change of heart about war as such. Non-official reports suggest between 1500 and 2500 service members were re-assigned to administrative or training duties due to selective conscientious objection.[21] Even if we assume that all of these re-assignments and discharges were genuine cases of selective conscientious objection, and allow that some units may have been degraded by loss of personnel, obviously the battle-effectiveness of the US military as a whole in the Gulf War was not greatly harmed.

Further support for the argument that selective conscientious objection is untenable can be found in the terrific problem the US intelligence community and the Special Operations Command have had since 9/11 in retaining highly-trained analysts and special operators who are offered more lucrative deals with private contractors. Here state agents are leaving government service for financial rather than moral reasons, but it is clear for all to see how damaging to the mission such high levels of turn-over can be when the door is open.[22] The US military has actually struggled to retain junior officers in all the services in the last decade even when lucrative private sector deals are not in the offing. Even if one accounts for the contingent stressors of serial deployments to Iraq and Afghanistan and the dissonance between ill-experienced senior officers (who had their field grade

20 As another chapter in this volume points out, Australia extended a right of selective conscientious objection to conscripts in the 1990s, although conscription itself had ended in the 1970s. The German experience will be interesting to observe as they begin to take an increasing role in certain military operations outside of Europe.

21 Keith Joseph, 'The Ethics of Selective Conscientious Objection', *Australian Defence Journal*, 160 (May, 2003), pp. 11-18.

22 Jeremy Scahill, 'The Secret US War in Pakistan', *The Nation*, 7 December 2009 http://www.thenation.com/article/secret-us-war-pakistan, accessed 30 September 2013.

experiences in the relatively peaceful 1980s) and combat-tested junior officers,[23] it is instructive to note that the attrition rate after the minimum five years of service for Marine and Navy officers graduating from the US Naval Academy (1984 to 2004) is significant, ranging from 15-21 percent. The rate climbs to between 47 and 59 percent after ten years of service for the period between 1980 and 1999.[24] The average attrition rate for West Point alumni after five years in the 1990s was 28 percent; the rate almost doubled for the first two classes to see combat in Afghanistan and Iraq, with 54 percent and 46 percent of the classes of 2000 and 2001, respectively, leaving at the end of their initial contracts.[25] One can reasonably assume that significant portions of those who leave the military at the end of their contracts would do so earlier if they could. It is beyond my competence to judge the next relevant question: whether these presumably higher attrition rates would be deleterious to the military's effectiveness. I have been told by various officers that US Defense Department constantly struggles to fill billets given *roughly* estimable future manpower needs, attrition rates, and recruitment classes. It is reasonable to assume that this task would be significantly harder, if the attrition rate variable were unbounded by mandatory terms of service.

In this chapter I urged a reframing of the issue of combatant equality as an empirical one, turning on the question whether selective conscientious objection is practically tenable for modern militaries. I argued that moral equality obtains if selective conscientious objection is untenable. While the available evidence does not offer a definitive answer, it seems to suggest that selective conscientious objection is not tenable. It is not out of the question that militaries might be reformed in order to accommodate selective conscientious objection.[26] One can conceive of technological change hastening this reform: further advances in robotics could take more and more of the actual fighting out of human hands and technological interfaces might require so little training that the incorporation of new recruits would not degrade the military's efficiency and efficacy. Under these conditions, there would be a strong argument against moral equality since militaries would not require iron obedience and multi-year service contracts and disaffected service members could leave as readily as attorneys can refuse unsavory clients. In this event, our attention would need to return to the question of reasonable epistemic expectations for service members in order to judge the question of moral equality.

23 Andrew Tilghman, 'The Army's Other Crisis', *Washington Monthly*, December. 2007 http://www.washingtonmonthly.com/features/2007/0712.tilghman.html, accessed 30 September 2013.

24 These statistics are communicated to me by the Director of the US Naval Academy's Internal Research Department, Capt. Glenn Gottschalk.

25 Lawrence Kolb, 'Reagan and the Draft', *Washington Times*, 16 May 2008, http://www.washingtontimes.com/news/2008/may/16/reagan-and-the-draft/, accessed 30 September 2013.

26 I owe this point to Bradley Strawser.

Chapter 6

Selective Conscientious Objection and the Just Society[1]

Dan Zupan

For the purposes of this chapter, I define a selective conscientious objector as a person in the armed forces who is not a pacifist and who otherwise willingly goes to war at his country's behest, but who refuses to fight a particular war because he has come to believe that the war is unjust. This essay explores whether it should be permissible for a soldier to claim selective conscientious objection status without fear of legal or other sanction. As it stands now in the American Armed Services, someone in uniform who refuses to go to war, for whatever reason, faces legal sanction, up to and including imprisonment. I argue that a just society should allow a person to claim selective conscientious objection without that person being subject to any negative sanctions.

I have written earlier that we never (or never should) give up our moral autonomy.[2] If I am ordered to do something immoral, I must refuse. In fact, in the military, we have a *jus in bello* obligation not only to refuse an illegal order, but to report the illegal order to higher authorities. If soldiers have the obligation to refuse illegal orders, then it seems that soldiers should also have the obligation to refuse to participate in a war whose entire premise is illegal. It is important not to conflate the illegal with the immoral, or vice versa. But for the purposes of this discussion, the sorts of illegal orders I will discuss are illegal *because* they are immoral. Their immoral nature is merely recognized and codified by statutory law. In like fashion, an illegal war is an immoral war – it is the crime of aggression, and it is a crime because it violates our most deeply held moral convictions. Walzer tells us that aggression 'is a singular and undifferentiated crime because, in all its forms, it challenges rights that are worth dying for.'[3]

---

1 If this essay has any merits, it is thanks to the critique I received from Roger Wertheimer, who pointed out some glaring inconsistencies and just plain wrong-headed thinking on my part. On the other hand, if the essay still falls short, only I am to blame.

2 Dan Zupan, 'A Presumption of the Moral Equality of Combatants: A Citizen-Soldier's Perspective', in Henry Shue and David Rodin (eds) *Just and Unjust Warriors: The Moral and Legal Status of Soldiers* (Oxford: Oxford University Press, 2007), pp. 214-25.

3 Michael Walzer, *Just and Unjust Wars: A Moral Argument with Historical Illustrations* (New York: Basic Books, 1977), p. 53.

Since it is the case that illegal orders are immoral ones (as I have defined them here), then the requirement to disobey and report them is a moral as well as legal obligation. Similarly, since an unjust war is immoral, the refusal to participate in it is a moral, not merely a legal obligation. As such, a society that demands that its soldiers refuse to obey immoral orders ought to be committed, by its own ideals, to support and defend any individual soldier's *legitimate*[4] claim of selective conscientious objection. That is, the soldier who believes a war to be unjust and refuses to participate should not face formal, official[5] negative sanctions. At the very least, we should recognize an inconsistency if society demands on one hand that he refuses to carry out an immoral order, say, to kill a non-combatant, yet, on the other hand coerces him, under threat of punishment, to participate in the mass murder of non-combatants in an unjust war.[6]

At this point we should address an odd asymmetry. If a soldier refuses an order to kill a non-combatant and has the moral courage to report the heinous order to his superiors, he is, or should be, lauded for his behaviour. But the selective conscientious objector who refuses to engage in the killing of countless innocents is the object of suspicion. Why is this? I offer here only one obvious reason – a more robust response would take us too far afield.

We are more inclined to support and praise the soldier who refuses to obey an illegal/immoral order because of the typical clarity of the situation. For example, if the lieutenant, in a fit of rage, orders the sergeant to bayonet a baby to 'inspire' villagers to provide information, the facts are immediate and obvious. Here, now, baby, kill. There is no ambiguity. The order is a clear assault against human decency, a clear violation of *jus in bello* proscriptions. But in the case of selective conscientious objection, the facts are not immediate and obvious. There is room for debate. The *jus ad bellum* assessments are far more complex and at a higher level of abstraction than the scenario about stabbing the baby. I submit that the relative moral uncertainty of the justice for a given war as compared to the relative moral certainty about the justice of a particular act in war must play an important role in any discussion of a soldier's obligations or permissions regarding his decisions to join or not to join the fray.

4 It should be implicit that we are considering the case of sincere selective conscientious objection, but I add this qualifier for emphasis. How we are to differentiate between the malingerer and the genuine selective conscientious objector is an important discussion but not one I engage in this paper.

5 He may well be ostracized by those of his mates who do not share his opinion, but we will insist that he suffers no punitive actions from his government. Moreover, his government should enact positive measures to protect him from the inevitable backlash attendant to his position.

6 War, whether just or unjust, does not logically require the killing of large numbers of civilians. However, empirical evidence concerning the civilian loss of life in any large-scale conflict demonstrates that this will inevitably be the case.

As hinted at above, many of the legitimate concerns regarding selective conscientious objection have to do with deeply entrenched ideas about the social contract, and subsequent considerations about the distinction between *jus in bello* (primarily a soldier's responsibility) and *jus ad bellum* (primarily a political responsibility). The United States Armed Forces are subservient to our elected officials. Soldiers are expected to fight when and where they are told to; they are expected, as well, never to start wars on their own initiative. Soldiers and officers swear to support and defend the Constitution. Their commitment to the Constitution binds them to obedience to their civilian leaders. It binds them to honour and obey the *jus ad bellum* decisions of the Republic.

Soldiers understand that fidelity to their oaths might require them to go to war, which entails that they (soldiers) might be harmed or killed and that they might have to harm or kill others. That is, in the case of the United States, unless the President directs them to fight an unconstitutional war, they must fight the wars they are told to fight. (It is not entirely clear what an unconstitutional war would be. If the President ordered the military to take over Congress, that would be unconstitutional, we must presume. On the other hand, the order to fight in an unjust war might not be unconstitutional; that is, there is nothing explicit in the Constitution that rules out aggressive wars). Thus, when soldiers and officers follow the directives of their elected leaders to fight a given war, they are fulfilling the conditions of their oath. And it is quite often assumed that if they are fulfilling their oaths, their actions are morally permissible. A stronger assumption is that fulfilling their oaths is morally obligatory. In fact, however, we should assume very little about whether they are morally bound to fulfil that oath or even whether they are morally permitted to fulfil it.

When a soldier makes their oath, they have consented to follow the mandates of their elected leaders. Some might argue that along with this understanding of the oath comes, among other things, the belief, in a democratic society, that the soldier will only be required to participate in a just war. In fact, however, the correlation between democratic societies and the justice of their wars is arbitrary at best. In spite of the rhetoric, democratic societies, or so-called democratic societies have, with great zeal, participated in or started many an unjust war.

Nonetheless, for the sake of this discussion and as a paean to the basic decency of the typical soldier, let us grant that the soldier presumes, given a constitutional, democratic society, replete with checks and balances, that they will be tasked to fight only just wars. This presumption is further justified by the fact that soldiers are inculcated with *jus in bello* requirements, as stated above. They are taught to discriminate between legitimate targets and non-combatants; apprised of the punishments under the Uniform Code of Military Justice that follow from transgressions; and so on.[7]

7 One need only conduct a brief Google search to discover the many soldiers and officers being tried and punished for various wartime crimes over the last decade.

Hence, although not articulated in such specific abstract terms, the reservations many soldiers and theorists harbour with respect to selective conscientious objection follows from the nature of the social compact and the ingrained tendency to presume the legitimacy of the decisions of the majority as embodied in our elected officials. Rawls counsels:

> If one views such disobedience as a political act addressed to the sense of justice of the community, then it seems reasonable, other things equal, to limit it to instances of substantial and clear injustice, and preferable to those which obstruct the path to removing other injustices. For this reason there is a presumption in favor of restricting civil disobedience to serious infringements of the first principle of justice, the principle of equal liberty… *Of course, it is not always easy to tell whether these principles are satisfied* (emphasis added).[8] … There are, to be sure, definite risks in the resort to civil disobedience. One reason for constitutional forms and their judicial interpretation is to establish a public reading of the political conception of justice and an explanation of the application of its principles to social questions Up to a certain point it is better that the law and its interpretation be settled than that it be settled rightly… Certainly in a state of near justice there is a presumption in favor of compliance in the absence of strong reasons to the contrary.[9]

Rawls expresses here a reservation concerning the conditions that must obtain before the discernments of the individual may rightfully claim parity (at least) with the law. However, he is not claiming that we should follow the law because the majority usually gets things right. Rather, he is making the pragmatic claim that, unless the law imposes harsh and widespread limitations on the liberties of others, it is most efficacious to obey the law and work for change from within the legal framework.[10]

In any case, resistance to selective conscientious objection often follows from a sentiment that collective wisdom, due process, checks and balances, etc., should trump individual conscience. Nonetheless, it is crucial that we recognize that nothing in a soldier's oath commits them to fighting unjust wars. When a soldier takes their oath, they are presumed to consent to follow orders, to go to war, to accept certain limitations on their personal liberties, etc. But it is implausible to suggest that they have also consented to participate in mass murder and wanton destruction. Just as Rawls views civil disobedience in the face of oppression as an inviolable right,[11] I view selective conscientious objection in the same light. I

8 John Rawls, *A Theory of Justice* (Cambridge, MA: The Belknap Press, 1971), p. 372.

9 Rawls, *A Theory of Justice*, p. 389. A crucial concern would be to establish what counts as 'strong reasons'. I defer this discussion for another project.

10 I owe this clarification to Roger Wertheimer.

11 Rawls, *A Theory of Justice*, pp. 363-82.

can see no version of a social contract, agreed upon for the purpose of securing justice, that could bind me knowingly to participate in the mass murder that is unjust war.

I have argued for the permissibility of selective conscientious objection. In the course of the argument, a concomitant obligation emerges to selectively object to participate in an unjust war. If a person sincerely believes a war to be unjust, they seem bound to refuse to participate in it. How could they knowingly engage in what they truly believe to be wholesale slaughter and consider themselves justified in doing so? The questions of guilt, admonition, of condemnation of soldiers who participate in wars that some consider unjust, hang heavily in the discourse about selective conscientious objection. If I believe a war to be unjust, for example, what attitude must I adopt towards my fellow soldiers who do not share my views? Am I to condemn them? Should society shun the soldiers? It might seem to be the case that our collective disapproval is demanded.

My response to the foregoing concerns follows from my conviction that the issue of selective conscientious objection is tied very closely to the issue of soldiers' ignorance with respect to *jus ad bellum* decisions. I have indicated that in a just, democratic society, our default with respect to decisions of elected officials is compliance. This preference for compliance follows from what we know about the basic structure of a minimally just society. Nothing I have said denies the inherent right of anyone to refuse to participate in unjust endeavours; on the other hand, my argument should not be construed as a complaint against soldiers who do not claim selective conscientious objection. This is so because of the difficulty anyone has in discerning the justice of any particular war.

Other Thoughts

I have been pressed by a lot of people to explain my decisions and actions throughout my military career, especially with respect to the second Gulf War, a war that I firmly believe to be an unjust, stupid, immoral war precipitated by one of the most thoroughly corrupt administrations ever to (mis)lead our country. I have been told something to the effect that anyone with a live conscience should have been troubled, morally, about serving in the military over the last 50 years. Much had been made of the great harm the military has caused during what some considered obviously unjust wars. Let me take up the case of one such war (use of military force) considered by some of my interlocutors to be unjust – the assault into Grenada. I use the Grenada case merely as a foil, a point of departure for discussion. If my brief depiction of the events is inaccurate, then I bid my readers to consider my description as an explanatory model.

Few of us in the military had even heard of Grenada. The units that conducted the initial airborne assault have an n-hour (notification time) sequence that requires them to be wheels up[12] within a very short time. There is no time for research, and soldiers' ignorance with respect to the justness of the cause actually is close to invincible. The opportunity to consider the justice of the cause, and hence, the possibility to legitimately consider opting for selective conscientious objection, simply does not obtain in these circumstances. This and other events over the years, along with formal study in ethics and the just war tradition, reinforce my belief that the distinction between *jus in bello* and *jus ad bellum* is a real distinction that characterizes the moral world of most soldiers. Even the most recent Iraq war highlights the difficulty in knowing the justice of any given war. Intelligent, well-meaning people still disagree on the moral status of the war. Given the epistemological challenge of anyone's knowing for certain the justice of particular wars, I feel strongly that the distinction (between *jus in* and *jus ad*) deserves a prominent place in any discussion about the moral status of soldiers and the case for selective conscientious objection.

I spent almost 27 years in the U.S. Army. My brothers and sisters in arms generally shared similar notions concerning our status as instruments of our government, instruments for defending the Constitution and resisting aggression. We acknowledge (along with most of the readership, I trust) that there are bad people, bad regimes, etc., and somebody has to be ready to oppose them. One reason I felt compelled to remain in the Army after my first term of service had ended had to do with the Cold War. I did feel a real obligation to be ready should our (and others') national survival be threatened. It may sound corny, but it is the truth: I did feel a duty. And it is critical to recognize that these were the prevailing sentiments of my brethren, men and women of the armed services. The very idea that we might be unjust warriors was never a part of our communal narrative. As junior officers, our horizons did not extend too far into time and space. Our moral preoccupations with respect to war concerned *jus in bello* scenarios. Especially in the aftermath of Vietnam, our military upbringing focused on just conduct in war. Recall that the pall of My Lai was a legacy that defined us in the minds of our civilian counterparts. Taunts of 'baby killers' still echoed in our collective consciousness. Small wonder that our focus was on matters *jus in bello*. It continues to be so. If we look through the prism of at least the US military's most recent history, we can understand that our primary concern is with justice in war. Moreover, we might be more inclined to conclude that this should be our primary concern. Certainly issues pertaining to the justice of war are profoundly important. But the moral arena wherein soldiers can have an impact and can influence decisions and actions for good or bad is located starkly within the discourse concerning the conduct of war. Given the ramifications on the ground, in immediate terms of human suffering, it is understandable that the soldier's fundamental moral concern revolves around just conduct in war. Moreover, since society judges his conduct on the same grounds as he judges himself (or since society *should* judge him and he *should judge himself*

12 These are airborne units. When we say we are required to be 'wheels up' within a particular time frame, we refer to the time of take-off of the first paratrooper-loaded aircraft.

on these grounds) the soldier's moral default to concern himself with justice in war versus justice for war appears practical and theoretically the appropriate domain of his concern. In fact, his commitment to just conduct in war is and ought to be, the fundamental object of his concentration. Otherwise, we might delude ourselves, and permit our soldiers the same delusion, that *jus ad bellum* is the only moral consideration that matters.

In our ruminations about selective conscientious objection, justice in war, justice for war, we might all be best served if we bear in mind that soldiers are committed to a profession that insists we obey *jus in bello* rules, indeed requires us to inculcate in each other, a commitment to fighting justly. At the same time, for most of us, it is not integral to our professional commitments to concern ourselves, at least in the initial stages of a war,[13] with the war's *jus ad bellum* status. For us to do so, from our perspective, would be a gross over-reach, a violation of one of our most sacred duties, the duty to obey our civilian leadership. And the real moral issue for us is not so much obedience per se, but what obedience represents. We recognize that one of the worst threats to any democracy would be an independent military. We fight in support of the Constitution and are very vigilant against any trends that might empower the military in inappropriate ways. Cynics may scoff at my assertions here, but I am relating the shared experience of those men and women who cut their teeth with me from the beginning of our careers over a quarter of a century ago. And our ethic is not unique; it shares its birth with that of the Republic.

So, whenever one accuses the military of having done something morally abhorrent (by being an instrument of what some may judge to be an unjust war), one should in the same sentence as the accusation recognize that we in the military are loathe to defy our civilian masters – not because we are lackeys devoid of moral courage; quite the opposite. Our commitment is to a way of life that expressly limits our power and authority; that requires us to follow the dictates of our political leaders. Doing so is, to us, an overt act of respect for the American people at whose pleasure we serve. It is, ultimately, a rejection of militarism.

Conclusion

My argument implies the legitimacy of the distinction between *jus in bello* and *jus ad bellum*. In fact, it provides justification for a *requirement* to recognize and embrace this distinction. What I have termed the logic of community[14] operates on a particular conception of the social contract theory as merely a conceptual tool that outlines the conditions that must obtain for people in a purported state of equal

13 I emphasize the initial phases of a war because it is during those first frenzied days and hours when time for research and reflection is not available. Typically, the soldier's ignorance is most operative at the beginning of most wars, before all the facts become known.

14 Zupan, 'A Presumption of the Moral Equality of Combatants'.

freedom to give over some of their rights (in the state of nature) to a communal process of decision-making. One of the more prominent rights we give over to the state is the right to use violence, when necessary, to protect the innocent and more generally to protect and uphold the institutions of liberty. If we require or permit soldiers to choose which wars they will fight (accepting the provisions for exceptions outlined above concerning unconstitutional wars), we require or permit them (however one wants to state it) to accept the social contract as specified throughout this paper, and to reject it. We say they must, and they must not, act in accordance with their oaths. This is an untenable situation; it requires that we accept a contradiction and ignore the cognitive dissonance that follows from it. So, from the perspective of justice as fairness, it may seem that we impose a stricture, a moral burden, which might not be consistent with an appropriate respect for soldiers and their predicament.

However, when we give over our individual rights to use violence, we do so, among other reasons, in order to secure a better state of justice. In a just, democratic society, we give the state the right and prerogative to use violence within a system of checks and balances, requiring due process, courts, trials, juries, Congressional approval, etc. That is, we do not surrender our right to use violence qua violence; rather, in giving over some of our rights to use violence, we are acting to prevent, as far as possible, the use of violence for evil ends. Hence, although the *jus in bello, jus ad bellum* distinction remains a credible moral criterion by which to judge politicians and soldiers (and for them to judge themselves), the judgments we make based upon this distinction are not unassailable. Just as the ignorance of the soldier is not invincible (with regard to *jus ad bellum*), neither is his innocence inviolable or his requirement to participate in any particular war absolute. A just society's responsibility to honour the individual choice for selective conscientious objection emerges as a moral requirement.

PART II
Case Studies in Selective Conscientious Objection

Chapter 7
Selective Conscientious Objection in Australia

Stephen Coleman and Nikki Coleman (with Richard Adams)

Ever since Australia became a country in 1901 there have been people unwilling to fight for 'Queen and Country' for various reasons; usually conscientious objection to warfare. In order to examine the way this issue has been handled by the military in Australia it is necessary to first look at how Australian law has dealt with the issues of both conscientious objection and selective conscientious objection. After examining how the law has dealt with these issues it will then be possible to see what the practical effect of this has been on Australian military operations since 1901, as well as what the implications of this might be for the Australian military in the future.

A History of Conscientious Objection and Selective Conscientious Objection in Australian Law

Those people who have an objection to warfare fall into two distinct groups: what might be termed 'general' conscientious objectors, who object to all forms of warfare, regardless of how noble the cause is; and selective conscientious objectors, who do not object to warfare in principle but object to specific wars for ethical, religious or even cultural reasons. A person might, for example, object to a particular war because: (a) they share a cultural background with those who would be regarded as 'the enemy' in that conflict, or (b) because they are of the belief that the war in question is not just and therefore it would be unethical for the soldier, and indeed for the country in general, to be involved in that conflict.

While the distinction is perhaps not as immediately obvious, those serving in the armed forces of the state can also be divided into two distinct groups: those who are conscripted into the military and have thus been forced to serve, and those who are volunteer members of the military. This distinction is particularly important when considering the issues of both general and selective conscientious objection, since under Australian law the rights of those conscripted into the military are somewhat different from the rights of those who have volunteered. In fact, in Australia the legal situation regarding the right of conscripted members of the military to conscientiously object to service is much clearer than similar rights for volunteer members of the military. This is perhaps a surprising situation given that volunteers make up 100 per cent of current serving Australian military personnel.

The Australian Defence Act (1903), which for simplicity will henceforth be referred to as the Defence Act, was the first legislation passed by any country in the world to grant 'total exemption from military service on the grounds of conscientious belief.'[1] Though it has been amended a number of times (as discussed below) this piece of legislation is still in force and it is immensely important with regard to the issue of conscientious objection, since the Defence Act is: (1) the piece of legislation which defines both the legal right to conscientiously object to military service and the limits on that right, and (2) the only current piece of legislation under which people can be compelled to serve in the Australian Defence Force (ADF). Under the Defence Act, persons conscripted to serve in the ADF may be exempted from service on the basis of either a universal or a specific, conscientious objection. However, while those who volunteer to serve in the ADF who come to acquire conscientious objection to military service may apply for discharge, as it currently stands the Defence Act does not allow volunteer members of the ADF to exercise a specific conscientious objection to a particular conflict in which Australia is involved.[2] The purpose of this paper is to explain both how this unusual situation came to be and what the implications of it are for the ADF in the future.

Under the 1903 Defence Act 'persons whose conscientious beliefs do not allow them to participate in war or warlike operations' are exempt from military service.[3] This includes people whose conscientious beliefs do not allow them to participate in 'a *particular* war or *particular* warlike operations', who are equally 'exempt from service in the Defence Force in time of war'.[4] Further, the Act provides that:

> s. 61B (1A) Persons whose conscientious beliefs do not allow them to engage in duties of a combatant nature (either generally or during a particular war or particular warlike operations) are not exempt from liability to serve in the Defence Force in time of war but are exempt from such duties while members of the Defence Force as long as those beliefs continue.

Significantly, the legislation does not require conscientious objection to reflect the doctrines of religious faith, acknowledging 'the atheist's conscience (to be) on a par with that of the Christian'.[5]

There were amendments made to the Act in 1910 when 'the reference to doctrines of religion as a test for conscientious objection that had appeared in the

1 Hugh Smith, 'Conscience, Law and the State: Australia's approach to Conscientious Objection since 1901', *Australian Journal of Politics and History*, 35 (1989), p. 13.

2 *Defence Act* (1903) s. 61C (c).

3 *Defence Act* (1903) s. 61A (1) (h).

4 *Defence Act* (1903) s. 61A (1) (i) (emphasis added).

5 Smith, 'Conscience, Law and the State', p. 16.

first Defence Act of 1903' was removed.[6] Another change was that conscientious objectors, though not required to bear arms in combat service, may well 'be required to perform non-combatant duties with the military forces in time of war'.[7]

World War I and World War II

For the first few decades of Australian independence, the questions raised by both conscientious objection and selective conscientious objection were essentially a non-issue in practice. The only war in which Australia became involved in this time was World War II. Many Australian men volunteered to serve in the armed forces and large Australian contingents were sent to the fighting in Europe. The government of the day wanted to introduce conscription in order to increase the number of troops who could be sent overseas to the conflict, but this move required an amendment to the law which only allowed conscripted forces to serve in direct (and local) defence of Australia. The issue was opened up to the vote of the Australian people through referenda in 1916 (narrowly defeated) and in 1917 (defeated by a wider margin).

As all Australian military personnel serving in World War I were volunteers, the legislation relating to conscientious objection was not tested, however the untested legislation was further amended in September 1939[8] when the term 'conscientious belief' was defined for the first time having the effect of making explicit that 'conscientious' included all moral belief, not just belief derived from religious doctrine. This made it possible for not only atheists to object to military service on moral grounds, but also allowed objection by members of religious denominations that were not pacifist in doctrine (e.g. a member of the Anglican Church could object on the grounds of conscience, even though the Anglican Church was supporting the war effort). World War II was the first time that the conscientious objection provisions of the Defence Act were tested since this war was the first one in which Australians were conscripted into military service. Compulsory military training was introduced in Australia at the outbreak of World War II but due to the nature of the Defence Act, only volunteers could be sent overseas for duty. Thus the Australian Expeditionary Force sent to serve in Europe consisted entirely of volunteers.

Japan's entry into the war in late 1941 brought with it a direct threat to Australia. During World War II well over half a million Australian men volunteered for military service and more than a quarter of a million men were

6 Hugh Smith, 'Conscientious Objection to Particular Wars: Australia's Experience During the Vietnam War 1965–1972', *War & Society*, 8 (1990), p. 119. Smith explains that, in this regard, Australia adopted a quite different course from the United States where 'legal argument over conscientious objection tended to focus on the question of religious tests for conscientious belief'.

7 Smith, 'Conscience, Law and the State', p. 16.

8 *Defence Bill* (No 2) 1939. Act No. 38 of 1939.

conscripted into the armed forces. Of course, only those men conscripted into the armed forces were eligible to claim exemption on grounds of conscience. When the amendments to allow conscientious objection were originally debated there was a fear that some members of society would use this 'loophole' to avoid their civil obligations, however this fear proved largely unfounded since few people applied for registration as conscientious objectors. 2791 applications for such exemption were received, of which 636 were rejected as not being genuine cases of conscience. Thus applications for conscientious objector status were received from about 1 per cent of those conscripted into service (and less than 0.3 per cent of Australians serving in the military during World War II). Given that Australia was directly threatened in World War II and that conscripts were only required to serve in immediate defence of Australia and its overseas territories (later extended somewhat to include the South West Pacific region), it can reasonably be assumed that all those who sought exemption from service on grounds of conscience in World War II were objecting to service in war in general, and thus that even if selective conscientious objection had been available at the time it is unlikely that any men would have applied for it.

National Service and the Vietnam War

In 1951, National Service was introduced into Australia (through the National Service Act), largely in response to the Cold War in Europe, the 1949 victory of communist forces in China and the outbreak of the Korean War in 1950. The number of people serving in the Australian military had been wound back since the end of the Second World War and National Service was designed to boost numbers of men who, although partly trained, could be called in to service should Australia be placed in a position of war. All eighteen-year old men were required to present for 176 days of military training, after which they were placed on the reserve lists for five years. This form of National Service was abandoned in 1957 by the Navy and Air Force, followed by the Army in 1959, largely because of a feeling that it was less important to have a significant number of semi-trained military personnel as it was to have a smaller number of skilled servicemen who had undertaken more extensive training. The training of large numbers of men unlikely to be called up to active service was also seen to be a large drain on the resources of the military, particularly of the Army. The legislation which had enabled this service had included provisions for conscientious objection which essentially repeated those contained in the Defence Act, as amended in 1939. The Defence Act had restricted the service of conscripts to within the Australian region, however no such limitations applied to those conscripted through the National Service Act. Various groups were granted an exemption from National Service, including Aboriginals and Torres Straight Islanders, volunteer members of the armed services, those who had joined the Citizen Military Forces (i.e. the military reserves), non-naturalized migrants, ministers of religion, theological students, the disabled, men who could prove that service would lead to undue hardship on their family, and men who had married before they were required to register,[9] although many of these exemptions were later removed, or altered from outright exemptions to deferments (though these deferments were sometimes indefinite).

Some significant questions about conscientious objection to military service arose in Australia in relation to National Service during the 1950s, but the issue really came to a head during the 1960s. Concerns about ongoing conflicts in South-East Asia – much closer to home than previous conflicts had been – saw a reintroduction of a form of conscription through an amendment to the National Service Act in 1964. All twenty year old males were now required to register for national service, though only a percentage of these men would actually be required to serve. Each year two ballots were conducted during which marbles, representing birthdays, were drawn out of a barrel. Registered men whose birthday was selected were eligible to be conscripted into the Army,[10] while all other men in that particular pool were informed that their requirement to serve in the armed forces had been indefinitely deferred. Those who were conscripted were required to serve two years in the Regular Army, followed by three years in the Citizen Military Forces (CMF, later formerly renamed the Army Reserve). In the period in which this National Service scheme operated (1965-1972), over 800,000 men registered for National Service, of whom nearly 64,000 were conscripted. The major reason that this new National Service scheme was controversial was that Australia became directly involved in the Vietnam War and it was obvious that National Servicemen would be deployed to Vietnam. This is the only conflict in which Australian conscripts have fought in anything other than the immediate defence of Australia; around 19,000 men, more than 35 per cent of the military personnel sent to the conflict by Australia, had been conscripted into service.

There were various ways in which a person might either avoid being sent to serve in Vietnam, or avoid serving in the armed forces altogether. A man who volunteered to serve in the CMF before his age group had been balloted was exempt from being drafted into the Regular Army. This was unattractive for many, since such volunteers were required to serve six years in the CMF; however by 1968 approximately 17,000 of the 35,000 members of the CMF were men who had taken this option. As Jeff Grey notes, men 'who exercised this option and then found that their birthdates had not been balloted into national service, became ineffective and avoided parades until finally discharged. As a result, the CMF was viewed by some sections of the community as a haven for draft dodgers.'[11] Another alternative was to attempt to avoid registering for the draft in the first place, or if that failed, to leave the country. From June 1968, men turning twenty who

9 'Conscription' in Peter Dennis et al. (eds) *The Oxford Companion to Australian Military History* (2nd edn), (Sydney: Oxford University Press, 2008), p. 159.

10 All men conscripted under this scheme served in the Army. The Navy and the Air Force were (and still are) composed entirely of volunteers.

11 Jeff Grey, *A Military History of Australia* (3rd edn), (Port Melbourne VIC: Cambridge University Press, 2008), p. 240.

intended to travel overseas needed to first obtain government permission; airline and shipping companies were instructed to not issue tickets to men in this age group until such permission had been obtained. Thus any man in this age group who attempted to travel internationally on an Australian passport would inevitably be under close scrutiny. But not all international travel required the use of an Australian passport, since any Australian could travel to New Zealand without a passport, and many men were able to obtain British passports which allowed them to travel internationally without scrutiny. Even after being drafted into the Army there were ways that one might avoid having to serve. One of the most high profile Australians to be drafted into National Service was Australian Test cricketer Doug Walters, who wrote a little about his experiences many years later.

> When I first went in for the National Service, I did three months at Kapooka on basic training … In the first couple of weeks we found several of the boys weren't too impressed with playing soldiers. One bloke went berserk with a bayonet. He didn't hurt anyone, but after spending some time cooling his heels, he was out of the Army. Another fellow was in the room next door to us. He'd hop out of his bed every morning, pull the bed clothes down, and relieve himself on the sheets. After a week or so of this, he was discharged with 'uncontrollable bladder'.[12]

While Walters did complete his two years in the Army, he managed to find another means of not serving in Vietnam. After finishing the second stage of their training at Singleton all the conscripts, Private Walters included, were asked to nominate first, second and third preferences for postings to a battalion.

> My first choice was obviously the First Battalion at Holsworthy, 1 RAR, because it was in Sydney … There was a slightly ulterior motive in plumping for Holsworthy, too: members of 1 RAR at the time were in Vietnam. They were part of the first Australian Regiment to go to Vietnam, so I thought there was a good chance that outfit would not be called on again while I was with them … Despite my battalion being in Vietnam when I first joined the Army, their turn to go again came quicker than I expected, as two battalions at a time were now being sent off. About six weeks before I was due to leave the Service our battalion was told it was going to Vietnam. Fortunately for me, you had to have at least three months to serve, before they would send a National Serviceman.[13]

Apart from joining the CMF, avoiding registration for the draft, or any of the methods of avoiding service which Walters describes, there was one other major way of dealing with the issue of conscription, through the law allowing for conscientious objection. The relevant law, for the purposes of determinations of

---

12 Doug Walters, *The Doug Walters Story* (as told to Ken Laws), (Adelaide: Rigby, 1981), pp. 35-6.

13 Ibid. pp. 38, 44.

conscientious objection, was the National Service Act, since the Defence Act only permitted conscription into military service for conflicts which involved the threat of direct invasion of Australian soil, into which category the conflict in Vietnam did not fall. While the law at the time only permitted general conscientious objection, a number of people who applied for conscientious objector status clearly had objections of a more specific kind. Objections to national service on the basis of selective conscientious objection were considered by magistrates, and in a few cases were ruled to be valid. However, in all such cases the federal government appealed the decision to a higher court, where the appeal was invariably successful and selective conscientious objection was ruled to not be a valid reason for avoiding national service. The nation's highest courts consistently ruled that the National Service Act only allowed for complete exemption from military service for men who could demonstrate a total pacifism. Between 1965 and June 1971, rulings were handed down on the cases of 1012 men who had applied for total exemption from military service on conscientious grounds. Of these, 733 were granted total exemption, 142 were granted exemption from combat duties (but not from non-combat duties) and 137 men were refused any exemption.

The fact that Australia's highest courts refused to grant exemptions from conscription on the grounds of selective conscientious objection makes it very difficult to realistically assess how many men might actually have applied for exemption from service in the Vietnam War on these grounds. Knowing (or perhaps at least suspecting) that the courts were almost certain to refuse requests for exemption from National Service based on objections to the war in Vietnam specifically, rather than to war in general, it seems likely that most men who did in fact object to such service would have sought other means to avoid it: joining the CMF; dodging the draft by some means such as leaving the country, or; accepting National Service but then trying to find ways to force the Army to end that National Service, such as by one of the methods described by Doug Walters.

During the course of the Vietnam War, the Federal opposition (i.e. the Australian Labor Party) did make some attempts to amend the National Service Act in order to allow exemptions on the grounds of selective conscientious objection. In the words of the Labor member Jim Cairns (who was later to become Deputy Prime Minister), 'If conscience is to mean anything, it must be based on the right of the individual to say what he believes is right and wrong'. Cairns further claimed that it did not matter if that individual 'happens to be wrong or right according to my standards or those of the government, or whether his view accords with the Gallup poll'.[14] In opposing any amendments which would allow selective conscientious objection, the Liberal member and future Minister of Defence, Jim Killen argued that:

> If a person has a conscientious objection, be it based on religious grounds or on philosophical grounds, to participating in a war, either in a combatant or in a non-combatant capacity, I have the utmost respect for him. But we are not

14 Quoted in Smith, 'Conscientious Objection to Particular Wars', p. 125.

> arguing that point. What we are arguing now is the judgement to determine whether or not a particular war is right or wrong; whether or not it is a just or an unjust war. Who determines that? … (this) is no longer a judgement based on conscience but a political one.[15]

Unsurprisingly, these moves to amend the National Service Act so as to permit selective conscientious objection were defeated along party lines. It was not until the Labor party was eventually returned to power late in 1972 that any further steps could be taken. But by this time the process of withdrawing Australian combat troops from Vietnam was already well underway, with the last combat forces leaving Vietnam only a week after the election. So rather than amending the National Service Act the new government simply suspended registrations for national service, before passing new legislation to terminate the scheme entirely in 1973.[16]

After Vietnam, the 1991 Gulf War and Beyond

There were no further changes to relevant Australian legislation for many years after the Vietnam War, but this did not mean that the issue of conscientious objection was never discussed in that time. In 1982, Senator Michael Tate put forward a Private Member's Bill that would change the conscientious objection provisions of the National Service Act (which was dormant having been suspended by the 1973 legislation, but was technically still on the statute) so as to incorporate objections to specific conflicts. This bill was referred to the Senate Standing Committee on Constitutional and Legal Affairs, and although the bill did not proceed any further, it did have one very important consequence; the Senate Standing Committee endorsed the proposed changes and recommended that the conscription provisions of both the dormant National Service Act and, more importantly, the Defence Act, be altered so as to allow objections to service based on selective conscientious objection. While this recommendation was not acted on at the time, it was certainly taken into account the next time Australian legislators had cause to consider the issue of selective conscientious objection, after Australia's involvement in the 1991 Gulf War. But this time the question was not whether conscripts ought to be able to conscientiously object to particular wars, but rather whether volunteer members of the Australian Defence Force ought to enjoy this right as well.

Following the Iraqi invasion of Kuwait in 1990, Australia joined in the international commitment to enforce trade sanctions on Iraq in the hope of reaching a peaceful end to the situation. Accordingly, a significant number of permanent Australian Defence Force personnel, primarily from the Navy, were dispatched to the region of the Persian Gulf to assist in the trade blockade and in preparations for a possible war to force Iraq out of Kuwait. One member of the Australian Navy

---

15 *Australian House of Representatives Hansard*, 28 May 1968, p. 1619.

16 National Service Termination Act (1973).

who objected to being required to serve in these operations was Leading Seaman Terry Jones. His case is worthy of note, since rather than appealing for exemption from service in the Australian contingent to the Gulf War (in August 1990), Jones left his ship and went absent without leave (AWOL) and chose to make a public statement through Senator Jo Valentine of the Greens Party:

> I am not a coward and I would be prepared to fight for my country, but I am taking a political stand because this is not our war, we are just following the Americans. I am prepared to die to defend my country but not to protect United States oil lines.[17]

Terry Jones was ultimately court martialled and convicted of absenting himself without leave, demoted in rank, forfeited four days' pay, and received a 21 day suspended sentence. He was discharged from the Navy shortly after, at his request. However, his case might potentially have had a very different outcome if he had not 'gone public' and raised his concerns about this service in such a politically charged manner. At the time of the first Gulf War there were no clauses in any Australian legislation allowing for conscientious objection to particular wars or warlike operations. While it was later suggested that all members of the Defence Force had been given a choice about whether they wished to serve in the Gulf, it appears that at least some serving members of the Australian military were not given clear information about this potential for choice, as evidenced by statements made by Senator Valentine when raising this issue. Senator Valentine pointed out that Senator Ray, the Defence Minister, had stated that service within the Australian contingent to the Gulf was voluntary, however Terry Jones had been not been advised of the option of a transfer from his ship prior to his deployment.[18]

There are a number of reasons why people object to the idea of allowing selective conscientious objection by serving members of the Defence Forces. For example, some people object to it because of a view that allowing selective conscientious objection would cause significant bureaucratic problems for the military machine which has to deal with these conscientious objectors; others suggest that the actions of these people might undermine the morale of the Defence Force as a whole. If the Australian deployment to the Persian Gulf in 1990-91 is taken as a case study, then these sorts of objections seem, at least, overdrawn. During the lead up to the first Gulf War (1990-91) and the combat operations that eventually ensued, there were 1108 personnel serving with the Australian contingent in the Middle-East (947 Navy, 49 Army and 112 Air Force personnel). Of these only twelve people, all from the Naval contingent 'are known to have not volunteered for the Australian commitment to the Gulf War' and of those twelve 'it is likely that eight (those released at their commanders discretion) were released

17 Quoted in Ian Wing, 'Selective Conscientious Objection and the Australian Defence Force', *Australian Defence Force Journal*, 137 (1999), p. 34.

18 Ibid.

for compassionate reasons which were unrelated to (conscientious objection). This leaves a figure of four possible selective conscientious objectors from the total Australian contingent, or 0.36 per cent.'[19] Such small numbers applying to not serve due to their conscientious objection to a particular operation – especially an operation which generated significant opposition in Australia, as the Gulf War did – would seem to negate the arguments that allowing serving volunteer members to choose not to serve in particular conflicts due to conscientious objection will undermine the morale of the ranks or create undue bureaucratic difficulties in dealing with those selective conscientious objectors.

The involvement of Australia in the Gulf War of 1991 ignited debate on the issue of selective conscientious objection and this led to a 1992 proposal to amend the Defence Act so as to allow exemption from service for 'persons whose conscientious beliefs do not allow then to participate in a particular war or warlike operations'.[20] The proposed legislation also removed exemptions from conscription for females and Aboriginals, and formally repealed the National Service Act.[21] It must be noted however, that all of these legislative changes were only in relation to conscripts and not volunteers, despite the fact that since the end of the Vietnam War all members of the Australian Defence Force have been volunteers and not conscripts. During discussion of this bill there were suggestions that the Defence Act ought to extend the right of selective conscientious objection not only to conscripts, but also to those who have volunteered for service in the military. For example, while the Australian Democrats supported the general provisions of the bill the party was concerned that these provisions did not go far enough:

> the Democrats are particularly concerned that the bill does not give volunteer service personnel the right of conscientious objection to service in overseas conflicts. Many Australians including some of the Defence Force have strong moral objections to such service.[22]

The Democrats were supported in this concern by others in the Senate, including members of the Green Party. However, both the Labor government and the main opposition (the Liberal/National coalition) were opposed to any change in legislation which would grant a right of selective conscientious objection to Defence Force volunteers (in fact the Coalition actually opposed allowing selective conscientious objection for any person at all, even those conscripted into service). In speaking against the proposal to allow selective conscientious objection for volunteers, the Defence Minister, Senator Ray stated:

19 Ibid. pp. 34-5.

20 *Defence Act* (1903) s. 61A 1(i).

21 Wing, 'Selective Conscientious Objection and the Australian Defence Force', p. 35.

22 Senator Vicki Bourne, *Hansard (Senate)* 29 April 1992, pp. 1818-20, quoted in Wing, ibid.

> I have to say I draw the line at that. There is no way that this Government can support that proposal … when people sign up to go into the Australian Defence Forces, they do so in the knowledge that they could be involved in any number or any type of conflict.[23]

In the end, legislation to amend the Defence Act was passed in 1992 and 1993. The legislation was passed essentially as it was originally proposed: allowing selective conscientious objection for conscripts, but not for volunteers; removing exemptions for Aboriginals and women; and formally repealing the National Service Act.

Amendments to the Defence Act passed in 1992 and 1993 also provided for the establishment of Conscientious Objection Tribunals, to provide rulings for applications for exemption for service on the grounds of conscientious belief.[24] Previously rulings on the issue of conscientious objection to military service had been made by a magistrate, as a part of formal legal proceedings. The aim of moving to a tribunal was to make the process more informal and less adversarial; the tribunals are made up of three members, one of whom (the Chair) must be a judge or retired judge. The determinations of the Conscientious Objection Tribunals are final and binding;[25] however the rulings are subject to appeal before the Administrative Appeals Tribunal.[26] Whilst this system of review before a tribunal is a vast improvement on the previous system, it is important to note that objections to military service (both in general terms and in regard to specific conflicts) are made on moral grounds, however the current system of examining these moral concerns are made through legal channels, blurring the boundaries between legal issues and moral issues raised by the person called to military service.

Even though the legislation only provides for the use of Conscientious Objection Tribunals to be used in establishing the exemption status of conscripts, in practice Australia has moved toward the use of an informal system in dealing with issues of conscience by volunteer members of the Australian military. This is evident in the administrative manner in which those who objected to service in the Persian Gulf in 1990-91 were dealt with, though the case of Terry Jones is an obvious exception here, although this is probably largely due to him going public about his concerns, an 'unforgivable sin' in Australian military culture.

There have been no legislative changes in Australia in regard to conscientious objection since 1993, and there have been no public displays of dissent such as by Terry Jones in 1990. This leads to the conclusion that informal administrative processes giving serving military personnel an opportunity to apply for exemption

23 Senator Robert Ray, *Hansard (Senate)* 23 June 1992, pp. 4291-93, quoted in Wing, ibid.

24 'Conscription' in Peter Dennis et al. (eds) *The Oxford Companion to Australian Military History* (2nd edn), (Sydney: Oxford University Press, 2008) p. 155.

25 *Defence Act* (1903) s. 61CC(2) (as amended).

26 *Defence Act* (1903) s. 61CZB (as amended).

from service in conflicts or deployments to which they have a conscientious objection, must be working reasonably well. There have been controversial deployments by Australian service personnel since 1993 (such as to Afghanistan since 2001 and to Iraq from 2003). Other states, like the USA and the UK, have also been involved in these controversial operations and those states have seen some high profile cases where military personnel have been prosecuted for failing to obey orders to deploy; essentially those personnel have been prosecuted for attempting to assert a right to selective conscientious objection. If Australian personnel were not given the opportunity to apply for exemption from deployments which have been seen to be so controversial within the wider community, it is likely that there would have been more public cases of dissent like that of Terry Jones or similar cases which occurred in other countries.

Selective Conscientious Objection in Australia: The Present and Future

The current legal situation in Australia is that a person can only be conscripted in to military service under the auspices of the Defence Act, which means that people can only be conscripted in the event that Australia is under direct military threat. Despite this, the act allows any person conscripted into military service to appeal against the requirement to serve on the basis of either a general conscientious objection, or selective conscientious objection to the particular war or warlike activity for which that person is being conscripted. In the event that a person does raise a conscientious objection to service, this appeal will be considered by a three person Conscientious Objection Tribunal, with this tribunal required to make a decision about a person's moral objection in accordance with the civil legal standard, based on 'the balance of probabilities'.

Those who have volunteered for military service, on the other hand, are not able to officially object to any particular war or warlike operation on the grounds of conscience. In practice though, the Australian Defence Force has tended to deal with such objections through an unofficial administrative process which has allowed those with moral objections to specific deployments or operations to obtain an internal transfer and thus avoid serving in the operation to which they have a moral objection. The Australian Defence Force could, of course, institute official legal proceedings against a Defence Force member who refuses, on moral grounds, to serve in a specific operation. However, it has been very rare for the Australian Defence Force to take this form of action; since the end of the Vietnam War it has only been resorted to in the case of Terry Jones, where the publicity surrounding the case perhaps left the Defence Force with little option but to take official legal action.

While volunteer members of the Australian Defence Force do not have access to selective conscientious objection, it is at least theoretically possible for a volunteer member of the Defence Force to access something like the general conscientious objection provisions open to conscripts. While the circumstances in which a serving member of the Defence Force comes to espouse total pacifism would admittedly be unusual, it is of course possible that a serving member might, for example, join a pacifistic religious organization and thus come to be a conscientious objector. Were this to be the case then that serviceperson could apply for a discharge from the military through the usual administrative procedures. However, even in cases where a serving member of the Defence Force comes to espouse total pacifism there might be difficulties for the member in obtaining a total discharge from the military: the member may not have completed their minimum period of engagement after enlistment, or not have completed a contracted period of service after reenlistment,[27] the member might have a 'return of service obligation' which has not yet been fulfilled,[28] it might be that the member possesses a critical skill which is in short supply within the Defence Force and thus that their discharge from the military would cause capability problems for the Defence Force, and so on.

One area that is yet to be fully tested legally is with regard to the rights and responsibilities of members of the Australian Defence Force Reserve, which is an important issues since reservists comprise a third of the total current membership of the Australian Defence Force. Due to changes made to the Defence Act in 1987, reservists can be called up to serve full time in situations less than in time of war or defence emergency. If called up in order to serve in a particular, and controversial, conflict, it is unclear if a reservist who objected to service in that conflict would be able to claim an exemption from such service on the grounds of selective conscientious objection, in the manner in which a conscript would be entitled to do, or whether reservists would be deemed equivalent to full-time members of the Australian Defence Force, and thus only able to claim conscientious objector status on the grounds of total pacifism. However, since those in the Reserve have not been conscripted into service, but have at some point volunteered, it seems most likely that Reserve personnel would face the same situation as full-time members of the Defence Force, in that they would not be able to object to service on the grounds of selective conscientious objection, although in such cases an unofficial administrative solution would most likely be found.

So to summarize, the current situation in Australia is that conscripted personnel have access to both general and selective conscientious objection, but

27 When members of the military reenlist after an initial period of service, it is common for this to be in the form of a contract wherein the re-enlisting member agrees to serve for a certain period of time, such as five more years of service.

28 There are a number of ways in which a member of the Defence Force might acquire a 'return of service obligation' (ROSO), but the most common is when the Defence Force pays for that member to complete a significant external education program, such as a full-time University degree. The usual policy in such cases is for the Defence member to agree to a ROSO which is equivalent to the length of that educational program plus one year. Thus if the Defence Force had paid for a member to complete a three year full-time University degree, then the ROSO for that degree would be four years of service after completion of the degree.

those who have volunteered to serve in the military only have access to general conscientious objection and cannot lodge an official conscientious objection to a specific conflict. However, the unofficial policy of the Australian military has been to seek an administrative solution in those cases where military personnel express concerns of conscience about involvement in specific conflicts.

Does it Really Matter?

Allowing access to selective conscientious objection for those conscripted into military service means that such persons will be able to make a moral decision about their involvement in a particular conflict based on the most up-to-date information available. Even more importantly, the clauses in the Defence Act which allow for selective conscientious objection for those conscripted into military service mean that there is a realistic chance that such a person's moral decision to not participate in a specific conflict will be respected under the law. For those who volunteer for military service, on the other hand, there are no such guarantees. Under the law as it stands, any person who volunteers for military service seems to be forced to give up any possible moral rights to object to service in particular conflicts at the time they volunteer. It is possible, even likely, that a person who volunteers for military service will learn a great deal about the morality of war after they enlist and thus this person might come to understand the moral importance of things such as the ideas expressed in just war theory with regard to *jus ad bellum* (i.e. when it is morally just to go to war) and *jus in bello* (i.e. how wars can be fought in a morally just manner). But if a military volunteer learns about such ideas after they have enlisted, even if they come to the firm conclusion that the war for which they have volunteered is actually immoral, then they no longer have any official recourse to avoid serving in such a war. An unofficial administrative policy which attempts – and only attempts – to allow military volunteers to avoid serving in a conflict which they believe to be immoral falls a long way short of the legal provisions which guarantee conscripts access to selective conscientious objection.

Some writers, such as Michael Walzer, argue that there is a more or less complete separation between *jus ad bellum* and *jus in bello*. Walzer claims that issues of *jus ad bellum* are solely the concern of a state's leadership and that military personnel need only to be concerned about issues of *jus in bello*.[29] Thus members of the military do nothing morally wrong if they fight on the unjust side of a war, provided that they actually fight the war in accordance with the rules of *jus in bello*. Other writers, such as Brian Orend, argue against this complete

29 Walzer claims, for example, that 'We draw a line between the war itself, for which soldiers are not responsible, and the conduct of the war, for which they are responsible, at least within their own sphere of activity. Generals may well straddle the line, but that only suggests that we know pretty well where it should be drawn.' *Just and Unjust Wars* (New York: Basic, 1977), pp. 38-9.

separation and suggest that *jus ad bellum* and *jus in bello* are inextricably linked to each other. If this is the case then it is difficult to see how military personnel who are engaged in fighting an unjust war can be totally excused from responsibility for fighting in such a war. As Orend suggests, we may consider such persons to be like minor accomplices to a major crime;[30] certainly not fully responsible for the *jus ad bellum* violations involved in engaging in the war, but nonetheless still not totally blameless. However, such arguments are entirely about what is moral in international armed conflict, not what is legal. In terms of international law there is certainly a separation between *jus ad bellum* and *jus in bello*; political leaders are held responsible for any *jus ad bellum* violations in a conflict while military personnel are only held responsible for *jus in bello* violations in that conflict. This means that it is perfectly possible for a particular conflict to be justified under international law, yet still be morally suspect, or for a conflict to be completely morally acceptable yet but problematic under international law.[31]

For all that it is recognized within the law, conscientious objection is NOT a view about what is or is not legal; it is a view about what is moral. It is perfectly possible, and logically consistent, for a person to be convinced that a particular action is totally legal and yet utterly immoral. Suppose someone was to argue that if military volunteers are given legal orders to participate in a particular conflict, then they simply ought to follow those orders and thus that there is no need for military volunteers to be able to express their selective conscientious objection to service in any conflict. On this view, even if the conflict is patently illegal, this is not a concern of the military personnel fighting in that conflict given the separation between *jus ad bellum* and *jus in bello* in international law. However any person expounding such an argument would have totally missed the point of conscientious objection; it is an objection based on CONSCIENCE, not an objection based on law. To suggest that it is morally reasonable to restrict the access of volunteer members of the military to selective conscientious objection seems to assume either: (1) that the state will never be involved in any unjust wars, a highly questionable assumption at best given that politicians seem to focus their attention on what is legal in terms of international conflict, rather than on what is moral, or (2) that there is total moral separation between the concerns of *jus ad bellum* and the concerns of *jus in bello* AND that military personnel will be convinced there is nothing wrong with following orders to fight in an immoral war, regardless of whether or not that war is legal under international law. If Australian volunteer military personnel are not able to express a moral objection to being ordered to fight in a war they consider morally unjustified, then it seems one must

30 Brian Orend, *The Morality of War* (Toronto: Broadview, 2006), p. 109.

31 For discussion of some of these issues see Stephen Coleman, *Military Ethics: An Introduction with Case Studies* (New York: Oxford University Press, 2012), especially chapters 4 and 5.

take seriously the argument made by Peter Tramel,[32] that it is always wrong to volunteer for military service, since his argument was based around exactly this idea. Establishing a clear, but formalized, administrative procedure to deal with selective conscientious objectors seems a reasonable approach; the model of the Conscientious Objection Tribunal could be a good one to follow here, since this approach may well be the best way to provide a legal forum in which moral concerns can be heard. If the percentage of selective conscientious objectors from the 1991 Gulf War is any guide, then managing such tribunals should not be too onerous for the Australian Defence Force.

Perhaps the biggest problem with the issue of selective conscientious objection in Australia is the fact that such an objection to military service is available to conscripts yet not to volunteer members of the military. A slightly fictional example seems the best way to throw this problem into stark relief. Imagine that Australia had conscripted people into military service in order to fight in the 2003 Iraq War. Those drafted into military service would have heard all the arguments about the justification for this war and would be able to take these things into account when determining whether or not to apply for status as conscientious objectors. But those who volunteered for military service would not be able to make such an argument, since their decision to volunteer – which may have been made when the world looked very different, before the September 11th attacks – is apparently taken by the Australian government to amount to fully informed consent to be deployed anywhere for any purpose. It is morally objectionable that those who have volunteered to serve their country and to risk life and limb in the process, should be thought to have less rights than those who are compelled into the same service.

32 See Peter Tramel, 'Immoral Military Service and Selective Conscientious Objection', conference paper presented at the Annual Conference of the *International Society for Military Ethics*, San Diego, January 2008. http://isme.tamu.edu/ISME08/Tramel08.html, accessed 30 September 2013.

Chapter 8
Conscientious Objection to Military Service in Britain[1]

Stephen Deakin

> Notoriously, we are dealing here with the most difficult if not the most insoluble of all questions, because the human conscience obeys no tribunal. It is its own tribunal. …We do not desire to have complications in the Army and to have to deal with persons who have scruples of the character which have been described. That would be a matter of great inconvenience.[2]

> The rights which have been granted in this war to the conscientious objectors are well-known and are a definite part of British policy. Anything in the nature of persecution, victimization or man hunting is odious to the British people.[3]

Introduction

Britain has a rich military history. Since the beginning of the First World War in 1914 the British military has been almost continually involved in conflict in some form or other – as they were in the previous century. Indeed, in most years since 1914 members of the British military have killed their enemies and many have died themselves on active service around the world. This rich and long military history has, inevitably, seen British forces involved in activities that have resulted in ethical controversy. Examples of this include two world wars, numerous counter insurgency campaigns, the area bombing of German cities in World War Two and the invasions of Suez and Iraq.

Despite such ethical controversy there has been very little conscientious objection to military service by civilian conscripts or by serving members of the military in Britain. Rarely has anyone refused to be conscripted, rarely have military personnel claimed that their orders are immoral ones that they must

1 Opinions expressed in this paper are the author's personal ones and they are not necessarily those of the Royal Military Academy Sandhurst, the British Army, the Ministry of Defence or the British Government. I am grateful for the helpful comments from the editors of this volume and also from Nigel de Lee.

2 Mr. Hore-Belisha debate on the Bill that became The Military Training Act 1939 that led to conscription during the Second World War and afterwards, House of Commons, *Hansard*, 18 May 1939.

3 Winston Churchill, House of Commons, 20 March 1941, *Hansard*, Vol. 307 Col. 284.

disobey. Instead, over the years since 1914, there has been much consensus about British defence policy. Almost all those involved in Britain's defence policy have served their country without protest and without conscientious objection. This is not to suggest that all those involved have been automata – disobedience to poor or objectionable orders by military personnel has been known.[4] However, in Britain conscientious objection to military service has been a rare event.

Conscientious Objection amongst Conscripts

Conscientious objection in Britain since 1914 is at its most obvious, and has attracted the most attention, during periods of national conscription. There is controversy when the state insists that eligible civilian citizens join the military to fight on behalf of their community and some refuse to do so. However, in Britain the very great majority of conscripted civilians have allowed themselves to be pressed into military service without protest.

Compulsory military service was a part of life for Britons from Anglo Saxon times and earlier. The Civil War was partly a war between conscript armies. The victories at Trafalgar and Waterloo were won on the backs of conscripts. There was some exemption on grounds of conscience from conscripted service for Quakers in the Militia Acts of the 1700s. The Nineteenth Century saw Britain avoid large wars and enthusiasm for conscription waned. Briton hid behind its large volunteer manned navy.[5] By the late Nineteenth Century Britain was not in favour of conscription and she placed a heavy emphasis on voluntarism.

At the beginning of the First World War in 1914 Britain was the only major European power to rely on voluntary forces. The heavy casualties soon incurred led to pressure for conscription. For the first time in British history national conscription was introduced by The Military Service Act 1916. This Act was very controversial and its passage led to heated debates in Parliament and in the country: the Cabinet was divided. Prime Minister Asquith, who steered the Bill through Parliament, had to use much political skill to gain acceptance for the measure and as a consequence his justification for it and its practical outworking were somewhat ambiguous.[6]

The Act allowed an exception to conscripted military service on grounds of conscience which its critics believed weakened Britain and encouraged cowardice and lack of patriotism. Its, mainly liberal, supporters maintained that there was a

---

4 Nigel de Lee, 'British Approaches to Military Obedience: Pragmatism, Operational Necessity and Moral Dilemmas', *Professional Ethics*, 10/2-4 (2002). See also Edward R Cain, 'Conscientious Objection in France, Britain and the United States', *Comparative Politics*, 2/2 (1970), pp. 275-307.

5 R J Q Adams and Philip P Poirer, *The Conscription Controversy in Great Britain 1900-18* (London: Macmillan 1987), pp. 24-5.

6 John Rae, *Conscience and Politics* (London: Oxford University Press, 1970), p. 48.

right to individual conscientious objection that the law must recognize. The Act did not, perhaps could not, define 'conscientious objection'; an omission that lay at the heart of difficulties with the Act and indeed the whole concept.

> Any certificate of exemption may be absolute, conditional, or temporary, as the authority by whom it was granted think best suited to the case, and *also* in the case of an application on conscientious grounds, may take the form of an exemption from combatant service only, or may be conditional on the applicant being engaged in some work which in the opinion of the tribunal dealing with the case is of national importance.[7]

This formal acknowledgement of the possibility of conscientious objection to military service has shaped British thinking and policy in this area ever since.

Those civilians who did not wish to be conscripted under the Military Service Act 1916 could appeal to a Military Service Tribunal. Over 5 million British men were in the military by the end of the First World War and approximately half this number had been conscripted – although some of these may eventually have volunteered anyway. Only 16,000 were conscientious objectors who refused to be conscripted. Many of these nevertheless worked in unarmed, uniformed military support roles; others worked in important civilian jobs. Others, 985,[8] refused to do anything to aid the war effort and were court-martialled and jailed for this. Often imprisoned with 'hard labour', such prisoners had to undergo an initial 28 day period of solitary confinement and further periods if they disobeyed the prison authorities.[9] Nine died of illness.[10]

Many of these objectors were highly educated, principled and actively religious or political; some were well connected socially and they and their supporters were able to publicize their views effectively. Eventually, after much controversy and lobbying of the Cabinet, a fudged solution was found whereby those who were deemed too ill to be imprisoned were released. Several of these imprisoned conscientious objectors went on to hold high positions in public service and four became members of the House of Lords,[11] whilst in 1940 six Members of Parliament had been conscientious objectors in World War One.[12] Official policy in the interwar period, 1919-1939, towards the possible reintroduction of conscription, contained an exception for conscientious objectors.[13]

7 *Military Service (No.2) Act* 1916, 2 (3).

8 Rae, *Conscience*, p. 201.

9 Ibid. p. 203.

10 Ibid. p. 226.

11 Ibid. p. 238.

12 Ibid. p. 243.

13 Ibid. p. 241. See also, Denis Hayes, *Conscription Conflict* (London: Sheppard Press, 1949), p. 373.

The political establishment gradually gave way on the issue of conscientious objection so much so that Prime Minister Chamberlain said in a speech to Parliament in 1939 about those who refused to serve in any capacity because of conscientious objection:

> We learned something about this in the Great War and I think we found that it was both a useless and exasperating waste of time and effort to force such people to act in a manner which was contrary to their principles.[14]

Chamberlain's sentiments exemplify British attitudes to conscientious objection by civilians. The Military Training Act 1939 was the initial legal instrument to provide conscription in Britain during the Second World War. It was much more accommodating towards conscientious objectors than the earlier legislation in 1916 and it again allowed civilian conscientious objectors to escape conscription to military service. It also allowed that serving soldiers could be discharged on conscientious objection grounds if they had committed an offence for that reason.[15] Surveying the developments in this area an authoritative study concludes, 'these provisions enabled those affected by the Act to exercise freedom of conscience to a degree unrivalled in any other country.'[16] However, around 6,500 people were still imprisoned for conscientious objection offences between 1940 and 1945.[17]

During the Second World War there are few known instances of conscripts refusing to fight on conscience grounds. One Second World War conscript who, famously, did refuse an order to fight was William Douglas Home, the brother of a future British prime minister. As his autobiography makes clear Douglas Home held at times both political and conscientious objector views against the conduct of the war and he did not fit in the Army.[18] He believed that the Allied policy of insisting on a German unconditional surrender led to many more lives being lost than would the negotiated settlement to the war that he favoured. Captain Douglas Home was ordered to take part in an attack on the French town of Le Havre in 1944. He refused to obey the order for fear of causing unnecessary civilian deaths.[19] Douglas Home maintained that the attack on Le Havre was a war crime; since it would, he believed, be both a disproportionate and an indiscriminate attack. He also argued that, as a result of his protest, in future attacks at Dunkirk and Calais civilians were let out of the towns before the battle began.[20]

The reaction of Douglas Home's senior officer, a Brigadier, is a revealing one. The Brigadier told Douglas Home that he had great respect for his views and in his

14 Prime Minister Chamberlain quoted in Hayes, *Conscription*, pp. 242-3.

15 Ibid. p. 244.

16 Ibid. pp. 244-5.

17 Caroline Moorehead, *Troublesome People* (London: Hamish Hamilton, 1987), p. 194.

18 William Douglas Home, *Sins of Commission* (Salisbury: Michael Russell, 1985).

19 Ibid. p. 60.

20 Ibid., p. 60.

official report on the matter he wrote that Douglas Home was a danger to morale and should not have contact with fighting soldiers in future.[21] The official reaction therefore to Douglas Home's conscientious disobedience of orders was to move him to some post away from the fighting or fighting troops. Douglas Home was tolerated by the Army and very probably his regiment would have successfully dealt internally with the problem he presented – as Douglas Home believed that it would.[22] This was not to be though once Douglas Home wrote, without official permission, a letter publicizing the incident to an English newspaper. He was eventually court martialled, cashiered and given one year's imprisonment with hard labour.

Conscription ended in 1945, but was reinstated in the form of National Service by The National Service Act 1948. Young men were required to join the military for up to two years and some of them saw combat in places such as Korea and Malaya in the 1950s. There was provision in the National Service arrangements for conscientious objection.[23] There were few National Service conscientious objectors.[24]

Conscientious Objection amongst All Volunteer Forces

Conscription and its peacetime counterpart National Service existed in Britain in the years 1916 to 1920 and from 1939 to the early 1960s. During other periods Britain has had all volunteer forces. Military recruitment themes in Britain have typically emphasized comradeship, challenge, education and sporting opportunities, travel and pay conditions. Volunteers who join the military are by definition self-motivated and they have promotion and careers to pursue. Military personnel are free to apply for discharge from their service employment, if they wish. When conscription, of the possibly unwilling, is no longer considered necessary or desirable, the case for conscientious objection is reduced.

When military personnel refuse a legal order on grounds of conscience they strike at the heart of what it is to be in the military and also at military effectiveness. Military personnel are part of an organized group that must put its members in danger of losing their lives to achieve its aims. Individuals often stay alive because they are protected by other members of the group whom they in turn protect. If individuals refuse to obey an order, they undermine the whole group and diminish its fighting power and effectiveness.

21 Ibid., p. 63.

22 Ibid., p. 61.

23 Len V. Scott, *Conscription and the Attlee Government* (Oxford: Clarendon, 1993), p. 101.

24 Bryan S. Johnson, *All Bull: The National Servicemen* (London: Allinson and Bury, 1973), p. 3; Gywn Harries-Jenkins, 'Britain: From Individual Conscience to Social Movement' in Charles C. Moskos and John Whitecray (eds), *The New Conscientious Objection* (London: OUP, 1993), pp. 67-79.

Incidents of conscientious objection in Britain since the First World War, outside the periods of conscription, are rare and they probably number fewer than a hundred or so cases. Courts martial for conscientious objection are also very rare. The established way for military personnel who find that they have conscientious objection to continued military service is to speak to their commanding officer. If they are considered to be a genuine case they are then discharged from the military on compassionate grounds. The evidence suggests that such resignations and discharges are mainly handled without acrimony and through established administrative procedures. It is of interest that protest resignations about aspects of defence management appear to have become more frequent in recent years and this perhaps indicates a decline in the consensus about defence matters that has typified Britain in the past.

In the 1920s Britain used Royal Air Force aeroplanes to bomb tribal villages and other civilian targets in Iraq. The British argued that this gave them control over wide areas at a lower cost than sending soldiers to hold territory. Officials argued that the bombing mainly affected the morale, the property and the way of life of the tribes' people whose casualties, it was claimed, were very low. Air Commodore Lionel Charlton CB, CMG, DSO, RAF, was second in command to the senior RAF officer in Iraq. Charlton was troubled by what he saw as the indiscriminate suffering of civilians as a result of the RAF bombing and in 1924 he offered his resignation from his post because of this. Charleton's autobiography, written in the third person, describes how he wrote a letter to his boss, 'stating that on grounds of conscience he felt he could no longer subscribe to the bombing policy' and that, accordingly, he could not carry out his orders and asked to be relieved of his duties.[25] At the same time some Members of Parliament were criticizing this air war and the issue had become a political one.

Charlton returned to Britain on leave. He was expecting an interview with the Chief of the Air Staff and head of the Royal Air Force, Trenchard, but no invitation came. Trenchard's official biographer notes that he did not want to make a martyr of Charleton and records the following dialogue when Charleton succeeded in gaining an interview with Trenchard. He was told, 'You resigned, and I accepted your resignation. There's nothing more to be said'. Charleton asked for an inquiry. Trenchard replied: 'An inquiry into what? Your conscience? Certainly not.'[26] Charleton records that this was a long and difficult interview, but that at the end of it he was given a promise that his future career would not be damaged by the matter.[27] Trenchard banned Charleton from serving in the Middle East whilst the bombing remained official policy and he was given a responsible job at the Air Ministry in England, although only a few years later he was forced to resign as being surplus to requirements. Charleton's conscientious objection was not a

---

25 Lionel Charlton, *Charlton* (London: Faber and Faber, 1931), pp. 279-80.

26 Andrew Boyle, *Trenchard* (London: Collins, 1962), p. 511.

27 Charlton, *Charlton*, p. 286.

refusal to serve in the military, but a refusal to carry out or to be associated with a particular policy, that of aerial bombing of tribal civilians.

The Suez conflict in 1956 was a controversial campaign conducted without a national consensus in Britain. The military carried out the operation with enthusiasm, but it is known that many senior military officers and civilian advisers expressed their reservations about the aim of the campaign. These included Admiral Mountbatten, at the time First Sea Lord and Chief of the Naval Staff, and retired Field Marshals Montgomery and Slim.[28] It is also alleged that the military Chiefs of Staff were close to resigning because Prime Minister Eden pressed them to do a quick airborne assault which they did not think advisable.[29] The only known service member who may have refused to fight at Suez on conscience grounds was the pilot of a RAF Canberra bomber, Pilot Officer Dennis Kenyon. He was about to take off to attack Egypt when he collapsed the undercarriage of the aeroplane on the runway and thereby aborted the mission.

His Squadron Leader may have given him a kindly way out when his first comment to him was, 'What's the matter, Dennis? Did you push the wrong button?'[30] Later he told his leader that he had deliberately retracted the undercarriage because, 'I did not altogether approve of what we were doing in Egypt'.[31] He said that he had had not eaten for days before the incident and that he had thought about suicide afterwards. At his court martial the prosecution alleged that he had pressed the wrong button because he was too afraid to fly the mission. Kenyon's defence was the same as the one originally offered to him by his boss.

> I have no political or religious views; I gave that reason merely because I was dreadfully worried about my tragic mistake. It was far better, I thought, to say that I had intentionally caused the Canberra damage rather than to say I had made a mistake and was incompetent.[32]

He was sentenced to a year's military detention and dismissed from the military.[33] It is clear from internal official documents that the political and military authorities feared that Kenyon's court martial, if he claimed conscientious objection grounds for his behaviour, would attract debate and controversy about the rights and wrongs of the campaign.[34]

28 Ray Fullick and Geoffrey Powell, *Suez: The Double War* (London: Hamish Hamilton, 1979), pp. 186-7.

29 Ibid.

30 'Cyprus: Grounded Bomber', *Time Magazine*, 11 February 1957.

31 Ibid.

32 Ibid.

33 Ibid.

34 Keith Kyle, *Suez* (London: Tauris, 2011), pp. 384-5.

Captain Michael Biggs, a British Army officer, resigned his commission in the 1980s on grounds of conscientious objection.[35] Biggs came from a military family, was commissioned from Sandhurst and he completed 17 years military service. His objection to continued military service was on spiritual but not religious grounds. Time spent as part of the Army's operation in Northern Ireland crystallized his thoughts and his desire to leave. The Army went to great lengths to keep him, sending him on a much prized university degree placement and giving him interesting jobs that did not involve fighting or training to fight, including one where he was not required to wear uniform. At his court martial he announced that 'If ordered to bear arms I would find it impossible to comply'.[36] The Army allowed him to resign.

The election of a 'New Labour' Government in 1997 led by Prime Minister Blair saw some radical changes to Britain's defence policy. New Labour's defence strategy was linked to a new internationalist foreign policy that saw Britain as a defender of human rights the world over. Not only was Blair's internationalism radical compared to the past, but also his government held progressive social views. The scene was set for some acts of conscientious objection from within the military.

One of New Labour's defence policy initiatives was to liberalize the treatment of homosexuals serving in the military. Whilst homosexual activity between consenting adults in private had been legalized in 1967, a ban remained on it within the military.[37] The military mostly had a de facto 'don't ask, don't tell' policy towards homosexual service people, but if someone's homosexual orientation became clear then action was often taken against them leading to a discharge and sometimes to a jail sentence. Opinion polls conducted within the military revealed a large majority supported the continued prohibition on known homosexuals serving in the British military. Amidst considerable controversy the policy was changed in 2000 so as to allow known homosexuals to serve.

A number of resignations over the issue were reported, particularly from senior ratings serving in the Royal Navy.[38] Two officers are known to have resigned on grounds of conscience over this issue. Brigadier Pat Lawless, the Deputy Commander of the Joint Helicopter Command, said that he could not reconcile his 'strongly held moral and military convictions as a soldier and a citizen with the government's decision to lift the ban … The decision – a very private matter – was

35 Details are in Moorehead, *Troublesome People*, pp. 318-20.

36 Ibid. p. 319.

37 See Stephen Deakin, 'The British Army and Homosexuality', in Hew Strachan (ed.), *The British Army, Manpower and Society into the Twenty-First Century* (London: Frank Cass, 2000), pp. 119-38.

38 'Lifting ban on gays in armed forces caused resignations, report reveals', *Daily Mail* 15 October 2007, http://www.dailymail.co.uk/news/article-487750/Lifting-ban-gays-armed-forces-caused-resignations-report-reveals.html, accessed 17 January 2012.

reached only after extensive reflection. I shall be very sad to leave'.[39] Commander Colin Douglas, a senior pilot and Director of Flying at the Royal Naval air station at Culdrose also resigned. He called the change of policy the 'final straw' and he argued that, 'There is plenty of sound military sense against lifting the ban, but the decision was taken for political and legal reasons'.[40]

The British involvement in the invasion of Iraq in 2003 led to considerable controversy. The British Government argued that it was a legitimate act, others argued that invading a sovereign country was illegal and prohibited by the United Nations Charter. Members of the British political establishment were divided about the issue. Meanwhile, there was a large protest march against the invasion in London, with perhaps between one and two million demonstrators. If the invasion was illegal, then British military personnel were in a difficult position and they might face legal proceedings accordingly. The head of the armed forces, The Chief of the Defence Staff, Admiral Boyce, wrote a letter to the Attorney General asking him to clarify the issue and to confirm in writing that the invasion was legal which he did. Admiral Boyce was seeking legal cover for deploying the British military. If anyone in the military did have conscientious objections to the war, then the Attorney General's opinion should have reassured them.

Attorney General Goldsmith has confirmed that he received a letter from Chief of the Defence Staff, Admiral Boyce, that asked for a 'yes or no answer' on the legality of war.[41] Members of the senior civil service also approached him for help.

> They were saying 'we are potentially at risk personally if we participate, or in the case of the Civil Service assist, in war if it turns out to be unlawful and therefore we want to know whether or not the Attorney's view is yes or no lawful'.[42]

No senior military figure resigned over the conflict, although a civil servant, Katherine Wilmhurst, who was a deputy to the Foreign Office's chief legal advisor, did so. Her resignation letter outlined her conscientious objection to the conflict,

> I cannot in conscience go along with advice – within the Office or to the public or Parliament – which asserts the legitimacy of military action without such a resolution, particularly since an unlawful use of force on such a scale amounts to

39 'Brigadier quits over gays in military' *BBC News*, 27 January 2000. http://news.bbc.co.uk/1/hi/uk/620178.stm, accessed 17 January 2012.

40 Tim Butcher, 'Pilot Resigns Over Lifting of Ban on Homosexuals', *The Daily Telegraph*, 28 January 2000.

41 Attorney General Lord Goldsmith, *Evidence to the Iraq Inquiry*, 27 January 2010, http://www.channel4.com/news/articles/uk/goldsmith+why+i+changed+iraq+war+advice/3518137.html, accessed 17 January 2013.

42 Ibid.

> the crime of aggression; nor can I agree with such action in circumstances which are so detrimental to the international order and the rule of law.[43]

There are few known cases of conscientious objection by military personnel over the policy of war with Iraq. This raises the question how such personnel are identified and counted. John McDonnell, a Labour Member of Parliament, argued in 2006 that over a thousand British military personnel had deserted since the start of the Iraq war, because they 'questioned the morality and legality of the occupation'.[44] He also suggested that, 'My understanding is there are a lot more seeking to avoid service, through different mechanisms'.[45] The Ministry of Defence replied that soldiers currently absent were considered to be 'absent without leave' and that they would have to be court martialled before being found guilty of deserting. Only one person had been found so guilty since 1989.[46] The Ministry of Defence maintained that 'absent without leave' numbers were unchanged and that, 'It often happens for family reasons and there is no evidence to suggest operational commitments contribute significantly to the figures'.[47] A solicitor who represented soldiers at courts martial suggested that the number of conscientious objectors who left Iraq was obscured since they were not counted as being conscientious objectors by the military.[48]

Ben Griffin was a trooper serving in the elite Special Air Service Regiment. Deployed to Iraq in 2004 he reached the conclusion that the conflict in Iraq was not a legal one. Out of loyalty to his soldier comrades he did not make a public protest whilst serving in Iraq, but on his return on leave in 2005 he asked to see his commanding officer. He told him that he would not return to Iraq, since he believed that the war was illegal and morally wrong and that politicians had lied about it to the people. He further alleged that American forces were routinely using too much force and that these were illegal tactics. Griffin thought that he would be court-martialled and sent to military prison. However, his commanding officer

> understood my point of view and his attitude was brilliant, in fact everyone was brilliant about it. I didn't know what was going to happen. I thought I might be charged or end up in Colchester [Britain's military prison] for refusing to soldier.[49]

---

43 Resignation letter by Elizabeth Wilmshurst, Deputy Legal Advisor to the Foreign Office http://news.bbc.co.uk/1/hi/uk_politics/4377605.stm, accessed 17 January 2013.

44 'At least 1,000 UK soldiers desert' *BBC News*, 28 May 2006. http://news.bbc.co.uk/1/hi/uk/5024104.stm, accessed 17 January 2013.

45 Ibid.

46 Ibid.

47 Ibid.

48 Ibid.

49 Sean Rayment, 'I didn't join the British Army to conduct American foreign policy', *The Daily Telegraph*, 12 March 2006, http://www.telegraph.co.uk/news/uknews/1512767/I-didnt-join-the-British-Army-to-conduct-American-foreign-policy.html, accessed 30 September 2013.

To his surprise he was allowed to leave the Army with his record unblemished and with a testimonial from his commanding officer that he was someone who was a 'balanced and honest soldier who possesses the strength and character to genuinely have the courage of his convictions'.[50] This is probably the first time an SAS soldier has refused to fight and left the Army on conscience grounds.[51]

Leading Aircraftsman Mohisin Khan was a Royal Air Force reservist medical assistant, who had served full time from 1999 to 2001 and who had been allowed to resign because he did not enjoy service life. However, he remained on the reserve list and could be recalled to the military because of his previous military service. He was recalled at the time of the Iraq war. He reported for duty and then said that for family and compassionate reasons he could not serve. This was rejected by the RAF, who maintained that he was going to be deployed not overseas but to help wounded personnel on their return to the UK. Khan went absent without leave, because he believed that he was to serve in Iraq. He argued that this was against his religious beliefs and that he did not want to fight against fellow Muslims.[52] His Flight Sergeant telephoned him the day after he had gone absent, tried to persuade him to return and suggested that he could discuss his concerns with the padre.[53] The RAF fined him nine days pay and seven days loss of privileges which he challenged in the High Court and lost. The High Court's opinion was that Khan was fully entitled to seek discharge on conscientious objection grounds, as was his right under the service regulations. However, he had never sought, or formally applied, to be discharged as a conscientious objector. Lord Justice Rix said,

> A volunteer cannot say that his conscience or religion has been interfered with by the state until he has made it clear in some appropriate and suitably formal way that he is no longer a volunteer.[54]

Another serviceman who refused to serve in Iraq was Flight Lieutenant Malcolm Kendall-Smith, a medical officer with the Royal Air Force. He had already served twice in the Middle East, but on the third occasion he refused to do so. He did not argue that he was a conscientious objector. He claimed that the war was illegal and that

---

50 Ibid.

51 Sean Rayment, 'SAS soldier quits Army in disgust at 'illegal' American tactics in Iraq', *The Daily Telegraph*, 12 March 2006, http://www.telegraph.co.uk/news/uknews/1512769/SAS-soldier-quits-Army-in-disgust-at-illegal-American-tactics-in-Iraq.html, accessed 17 January 2013.

52 'Muslim reservist loses Iraq case', *BBC News*, 7 October 2004 http://news.bbc.co.uk/1/hi/england/suffolk/3722812.stm; 'Muslim airman challenges RAF fine', BBC, 19 July 2004. http://news.bbc.co.uk/1/hi/england/suffolk/3907813.stm, both accessed 30 September 2013.

53 'War charge Muslim "feared hell"', *BBC News*, 12 September 2003, http://news.bbc.co.uk/1/hi/england/suffolk/3104210.stm, accessed 30 September 2013.

54 'Muslim airman challenges RAF fine', *BBC News*, 19 July 2004, http://news.bbc.co.uk/1/hi/england/suffolk/3907813.stm, accessed 30 September 2013.

he did not want to take part of an 'act of aggression' that was contrary to international law.[55] The prosecution argued that Kendall-Smith was not in a position to argue about the legal issues relating to the Iraq war and that the issue of concern was that he refused legal orders to undergo weapons training and equipment fitting prior to deployment. His case was weakened, since the war was already over when he refused to obey orders. It was further alleged that he was motivated by personal grievance that the RAF would not let him resign immediately from the service. For the prosecution this was not a moral or conscience issue, but a simple matter of a refusal to carry out straightforward orders. The Judge Advocate at the court martial decided that Kendall-Smith was not allowed to argue that the war was illegal. He accused Kendall-Smith of seeking to be a 'martyr'.[56] The sentence was intended to make an example of him and to serve as a warning to others in the forces. As the Judge Advocate argued,

> Obedience of orders is at the heart of any disciplined force…refusal to obey orders means that the force is not a disciplined force but a disorganized rabble. Those who wear the Queen's uniform cannot pick and choose which orders they will obey. Those who seek to do so must face the serious consequences.[57]

Kendall-Smith was sentenced to eight months imprisonment, had to pay £20,000 towards his defence costs and was dismissed from the service. It is understood that he was quietly released from jail early in his sentence.

Two British Army soldiers from 16 Air Assault Brigade reported that they had been sent home from the Middle East in 2003, because they were conscientious objectors who did not wish to fight in the Iraq war.[58] The soldiers feared court martial for their actions. A solicitor involved in their case said,

> I believe that these two soldiers have told their commanding officers that they will not fight in a war involving the death of innocent civilians…And as a result of that they have been returned to unit, what will happen now is a matter of conjecture.[59]

However, a Ministry of Defence spokesperson said that there was 'no record of anybody from 16 Air Assault Brigade returning on grounds of conscientiously

55 'RAF doctor jailed over Iraq refusal', *The Guardian*, 13 April 2006, http://www.guardian.co.uk/uk/2006/apr/13/military.iraq, accessed 30 September 2013.

56 'Prisoner of conscience: RAF doctor who refused Iraq service is jailed', *The Independent*, 14 April 2006, http://www.independent.co.uk/news/uk/crime/prisoner-of-conscience-raf-doctor-who-refused-iraq-service-is-jailed-474090.html, accessed 30 September 2013.

57 Ibid.

58 'Two soldiers sent home from Gulf war', *BBC News*, 31 March 2003, http://news.bbc.co.uk/1/hi/england/2900963.stm, accessed 30 September 2013.

59 Ibid.

objecting to the conflict in Iraq'.[60] The spokesperson confirmed that two soldiers had returned home and were now working in a reserve unit,

> We have no evidence that anybody has been sent back for refusing to fight... We do get soldiers sent back all the time from theatre for various reasons such as medical, welfare and disciplinary things that need to be dealt with back in the units.[61]

The soldiers' solicitor understood that they had not been charged with any offence and they were now working in a reserve party.[62]

Despite Britain's desire to deploy forces to Afghanistan without their being required to fire a shot in anger, they encountered heavy conflict almost from the beginning. The military found themselves involved in war fighting of an intensity that had not been seen by British forces since Korea in the early 1950s. This deployment challenged the military in a number of ways. Michael Lyons a Royal Navy sick berth attendant was ordered to deploy to Afghanistan and he was successfully court martialled in 2010 for refusing to practise pre-deployment rifle training, since he claimed that he had a 'moral objection' to carrying arms and would like a non-combat role.[63] Medical personnel may carry weapons for personal defence and for the defence of their patients. At his court martial Lyons said,

> My initial objections started with Afghanistan and I wanted to investigate the reasons why we were at war. At the time WikiLeaks came along and mentioned Iraq and Afghanistan. The reports said there had been some civilian casualties that nobody knew about and they were being covered up... After a lot of deliberation I decided I was a conscientious objector.[64]

Lyons also claimed that he had been instructed that he was not to treat injured Afghan civilians since it would be a waste of resources.[65] He applied to be considered a conscientious objector, but this was refused and he appealed to the Advisory Committee on Conscientious Objectors, which also refused his plea.

---

60 Ibid.

61 'When Soldiers will not fight', BBC News, 31 March 2003, http://news.bbc.co.uk/1/hi/uk/2903743.stm, accessed 30 September 2013.

62 Ibid.

63 'Court martial outlines war objector Michael Lyons' failed appeal', BBC, 1 December 2011, http://www.bbc.co.uk/news/uk-england-devon-15992275, accessed 27 September 2013.

64 'Navy medic detained for refusing training over WikiLeaks claims', *The Guardian*, 5 July 2011, http://www.guardian.co.uk/uk/2011/jul/05/navy-medic-afghanistan-rifle-training, accessed 17 September 2013.

65 Ibid.

At the hearing he said that he could not serve in Afghanistan after reading the Wikileaks and finding that he might be unable to treat anyone and added,

> If you're at a patrol base or forward operating base, it's likely you'll have to use your weapon and will have to turn civilians away who are in need of medical aid. … If more people in my position stood up, there would be a lot less innocent lives lost around the world.[66]

He was dismissed from the Navy and sentenced to seven months detention at a military correction facility. The Judge Advocate said that the court martial could not make a decision about his claim of conscientious objection,

> We are undecided whether your views are genuine or you grasped the straw of conscientious objection as you didn't wish to serve your term.[67]

Lyons complained that his human rights were being infringed. However, Lord Justice Toulson rejected his appeal saying, 'A person who voluntarily enters military service undertakes serious responsibilities potentially involving the lives and safety of others'.[68]

Lance Corporal Joe Glenton served in Afghanistan, but refused to return there and went absent without leave for two years during which time he went abroad. On his return he spoke at anti-war rallies and publicized a letter he wrote to the Prime Minister stating that Britain's commitment in Afghanistan was illegal and wrong. The military initially charged him with desertion, a crime that carries with it a sentence of up to ten years in jail plus five other offences, involving speaking to the media without permission, that could have led to further years in prison. Eventually the prosecution and defence agreed that Glenton was suffering from post-traumatic stress syndrome and the charge of desertion was dropped in favour of the lesser charge of being absent without leave. He was sentenced to nine months in jail and was likely to serve five months. The Judge Advocate said at the court martial, 'There are many soldiers in the army who have extremely unpleasant experiences, who watch friends die or suffer appalling injuries, but they have to return and do their duty'.[69] The Judge Advocate was also unconvinced about his anti-war beliefs, since Glenton had not raised them initially on his return from abroad.[70]

---

66 'Navy Afghan war objector Michael Lyons detained', *BBC News*, http://www.bbc.co.uk/news/uk-england-devon-14028167, accessed 26 September 2013.

67 Ibid.

68 Ibid.

69 'British soldier who spoke out against Afghan war jailed for going awol' *The Guardian* 5 March 2010, http://www.guardian.co.uk/uk/2010/mar/05/anti-war-soldier-jailed-afghanistan, accessed 29 September 2013.

70 Ibid.

Iraq and Afghanistan also led to protest resignations about Britain's defence policy from within the military. Illustrating the consensus in Britain about defence policy, this was protest about the manner in which the conflict was carried out, rather than the principle of British military involvement in these countries. Brigadier Ed Butler, a former commander of the SAS and an architect of Britain's initial strategy in Afghanistan, resigned his post of Chief of Joint Operations at the Permanent Joint Headquarters. In a statement he wrote that he must put his family first and that they were suffering due to his prolonged absences. However, he also criticized government policy, writing of the soldiers he had led: 'Their resourcefulness, commitment and ability to continually deliver operational success, within the well-known constraints and restraints, is both remarkable and humbling. We owe them a huge debt of gratitude.'[71]

Sources close to the Brigadier spoke of his frustration at the lack of resources for the Army.[72] Once retired, the Brigadier was more forthright:

> I think the wider campaign in Afghanistan, and this has been the case from the early days, has been insufficiently resourced to undertake a proper counter-insurgency. … I was the commander of British forces in 2006 and I made the case very clearly then that we had, only just about, enough resources to do what would be seen as a very steady state task.[73]

The commander of the Special Air Service was also reported to have resigned in protest at defence cuts in 2010.[74] Similarly, an SAS officer, Major Morley, the commander of D Squadron, 23 SAS, resigned at what he called in his resignation letter 'chronic underinvestment' in equipment, which was to blame, he believed, for the deaths of soldiers in Afghanistan.[75] The commander of 3 PARA, Lieutenant Colonel Stuart Tootal resigned after service in Afghanistan. His reasons for

---

71 Mathew Hickley, 'SAS hero quits with parting shot over army cuts', *Daily Mail*, 8 June 2008, http://www.dailymail.co.uk/news/article-1025032/SAS-hero-quits-parting-shot-army-cuts.html, accessed 28 September 2013.

72 Ibid.

73 'British military in Afghanistan "insufficiently resourced", says former commander', *The Daily Telegraph*, 22 July 2009. http://www.telegraph.co.uk/news/uknews/defence/5887092/British-military-in-Afghanistan-insufficiently-resourced-says-former-commander.html, accessed 29 September 2013.

74 'SAS commander quits Army amid claims defence cuts have hit morale', *The Daily Telegraph*, 27 November 2010, http://www.telegraph.co.uk/news/uknews/defence/8164888/SAS-commander-quits-Army-amid-claims-defence-cuts-have-hit-morale.html, accessed 28 September 2013.

75 'Exclusive: SAS chief quits over 'negligence that killed his troops', *The Daily Telegraph*, 31 October 2008, http://www.telegraph.co.uk/news/newstopics/onthefrontline/3332417/Exclusive-SAS-chief-quits-over-negligence-that-killed-his-troops.html, accessed 28 September 2013.

doing so were complex, but underfunding of the military was an important one.[76] Lieutenant Colonel Nick Henderson, Commanding Officer of the Coldstream Guards, resigned at the end of his tour of operational duty in Iraq. He believed that several of his soldiers had been killed due to their inadequate equipment.[77]

Conscientious objection to military service can take many forms, and resigning in protest is, in a highly consensual country like Britain, what military people do when they believe that they must make a conscientious stand.

Discussion of these Cases

The common view about British policy towards conscientious objection to military service is twofold. Firstly, it is accepted that Britain is tolerant, certainly much more so than many other countries, towards civilians who are determined to refuse conscripted military service on grounds of conscience. The evidence supports this view. Secondly, it is argued that Britain is hostile to selective conscientious objection by serving members of the military. Britain is sometimes thought of as having the same policies as countries such as the US and Israel. 'The US, UK and Israel have adopted a strong opposition to selective conscientious objection, especially for volunteers for military service.'[78] The evidence for this second assertion about British practice is not convincing. A better reading of the available evidence is that Britain accommodates selective conscientious objectors, if they behave reasonably and do not cause political embarrassment. It does this by the subtle use of administrative procedures.

Britain has not had problems that are likely to fuel conscientious objection, such as unpopular wars as the US had in Vietnam, nor has it had disputes over territory, such as the Israelis have with the West Bank. Only the Irish question has caused difficulties, in the recent past, with, for example, the 'Curragh Mutiny' in 1914[79] and, somewhat similarly, the Connaught Rangers mutiny in India in 1920.[80] At the same time Britain established in 1916 the principle that conscientious objection should be recognized legally. Some First World War conscientious objectors became influential martyrs who damaged the reputation of the British military and the British government. Britain has tried to avoid this and, since then, has mostly treated conscientious objectors with respect.

76 Stuart Tootal, *Danger Close*, (London: John Murray, 2009), p. 276.

77 Jack Fairweather, *A War of Chance* (London: Jonathan Cape, 2011), pp. 221-2.

78 Hyder Gulam and Mike O'Connor, 'Selective Conscientious Objection: The Court Martial of Flight Lieutenant Malcolm Kendall-Smith', *ADF Health*, 7 (October 2006), pp. 68-72. For support for this argument see, Paul Robinson, 'Integrity and Selective Conscientious Objection', *Journal of Military Ethics*, 8/1 (2009), pp. 34-47.

79 Ian Beckett (ed.), *The Army and the Curragh Incident, 1914* (London: Bodley Head, 1986).

80 Anthony Babington, *The Devil to Pay* (London: Leo Cooper, 1991).

It has done this by adopting where possible a low key approach to such issues and avoiding controversy. After all, in a liberal democracy such as Britain, members of the military are supposed to be apolitical and avoid the sort of controversy that, particularly, selective conscientious objection usually brings. One consequence of this is that official policy in this area and its outworking over the years is difficult to analyse. Critics of this policy stance argue that military personnel should be made much more aware of their ability to leave on conscientious objection grounds. Internal procedures for discharge on the grounds of conscientious objection by serving personnel exist in each British military service and they are administrative in nature; they are not readily available to the public. They do not exist in legislation, but in military regulations. A non-governmental organization obtained details of current procedures in 2007 by using the Freedom of Information Act. It concluded that cases are not necessarily recorded as cases of conscientious objection[81] and that the military do not highlight the possibility of conscientious objection to their personnel.[82]

Actual figures for those classified as conscientious objectors are tiny. In the period 1990-2000, 13 Royal Navy sailors and one Army soldier applied for conscientious objector discharge.[83] In the period 2001 to 2010 nine people, from the Navy and the Royal Air Force, applied for discharge on these grounds. Six of these were successful. No one from the Army has used this facility since 2001.[84] These figures are testimony to the lack of conscientious objection in Britain, or possibly, to the skill with which the Ministry of Defence handled any such issues. It was during this period that Britain formed part of a coalition to invade Iraq in 2003 in controversial circumstances. If ever conscientious objection was likely to occur among British volunteer military personnel, this was the occasion.

Once a military conscientious objector makes an application for discharge he continues to have all the responsibilities of a member of the Armed Forces and can still be deployed into a combat zone. There is an appeal process to the Advisory Committee on Conscientious Objectors, appointed by the Lord Chancellor, which makes a judgement of the applicant's conscience and case. The Committee dealt with 36 appeals between 1970 and 2001.[85] Leading Seaman Lyons' case was the first for 16 years to be heard by the Committee. It was only the 37th that the Committee had heard.

---

81 Forces Watch, see its publication by David Gee, *Informed Choice? – Armed Forces Recruitment Practice in the United Kingdom* (InformedChoice, 2007), p. 75.

82 Ibid.

83 UK Ministry of Defence, *Conscientious Objectors 1990-2000*, http://www.mod.uk/DefenceInternet/FreedomOfInformation/DisclosureLog/ConscientiousObjectors19902000.htm, accessed 29 September 2013.

84 'Conscientious objectors figures revealed', *The Independent*, 30 January 2011, http://www.independent.co.uk/news/uk/home-news/conscientious-objectors-figures-revealed-2198647.html?origin=internalSearch, accessed 30 September 2013.

85 *War Resisters International*, http://wri-irg.org/news/2007/uk2007-unreport-en.htm, accessed 29 January 2013.

In many ways this process does not differ greatly in substance from simply resigning from the military and giving, typically, 12 to 18 months' notice. The main difference, of course, is that members of the military cannot suddenly refuse to serve, because they believe that British forces should not take part in a particular conflict or undertake a particular action of which they disapprove.

However, the situation in reality is more complex than this. Britain learnt from the problems that it had with universal conscientious objectors in the First World War. Cases of selective conscientious objection, inevitably, publicize issues relating to a particular war that the government of the day and the military do not wish to be discussed in this context. The Ministry of Defence uses internal administrative procedures to deal with issues in this area and its processes do not lend themselves to identifying cases of conscientious objection. For example, soldiers 'absent without leave', some possibly for reasons of conscientious objection, have to be court martialled before they can be found guilty of deserting. This almost never happens. Only one person has been found so guilty since 1989.[86] Presumably, such soldiers are simply administratively discharged. It is clear from the evidence that the military tries, understandably, to avoid political controversy.

Whatever the formal policies about conscientious objection in Britain, they must also be considered within the context of the realities of military life and ethos. The military ethos is one of a close knit community that looks after its own and shares common, often intangible understandings about life. The ethos of the Army,[87] and the other services, is pragmatic, determined and impatient of philosophical reasoning. Those who fight depend on their comrades for support and to save their lives. No one wants a comrade alongside him who will not fight, for reasons of conscientious objection, or for any other.

The author has heard British officers who fought in Normandy in 1944 remark that they sometimes had posted to them battlefield casualty replacement officers who did not wish to fight. No action was taken against these. They were simply sent elsewhere to carry out non-combatant support roles. Similarly, Second World War Royal Air Force aircrew who refused to fly, for whatever reason, were moved away quickly, often labelled as having 'lack of moral fibre' and demoted to menial jobs if they were sergeant pilots. Officer pilots were asked to resign or given non-combat jobs.[88] A General argues that Army units continually and actively weeded out unsuitable recruits from front line combat units,

86 'At least 1,000 UK soldiers desert' *BBC News*, 28 May 2006. http://news.bbc.co.uk/1/hi/uk/5024104.stm, accessed 27 September 2013.

87 See Stephen Deakin, 'Education in an Ethos at the Royal Military Academy Sandhurst', in Paul Robinson, Nigel de Lee, and Don Carrick (eds), *Ethics Education in the Military* (Aldershot: Ashgate, 2008), pp. 15-29.

88 Neil Greenberg et al, 'Mental Healthcare in the United Kingdom Armed Forces', in Elspeth Cameron Ritchie (ed.), *Combat and Operational Behavioral Health* (Fort Detrick, Maryland: Borden Institute, 2011), p. 659. http://www.bordeninstitute.army.mil/published_volumes/combat_operational/CBM-ch41-final.pdf, accessed 28 September 2013.

> the Depot continued to churn out a number of men who for reasons of mentality, aptitude, intelligence or honesty would never have got near a D-Day battalion and getting rid of these was a time-consuming and never-ending business. Certainly there is no place on operations for a man who pilfers from his comrades or absents himself without leave.[89]

Tim Collins, Commanding Officer of the Royal Irish infantry regiment, which played an important role in the invasion of Iraq in 2003, writes, 'When one of my soldiers expressed doubt about crossing the border into Iraq in March 2003, I agreed it might be better for him to remain behind. I had already said that I wanted no man by my side who did not want to be there.'[90] HMS Coventry, a Royal Navy destroyer, was sent at no notice to join the conflict in the Falklands Islands in 1982. Her Captain asked each crew member whether they were prepared to go to war or whether they wanted to go home. Five of the crew swapped places with volunteer seamen from another ship that was returning to the UK. Coventry's Captain argued that, 'It was very important to take all the strong people and not have anyone who really ought to be at home.'[91]

Those in the military community look for total reliability and commitment in each other. Without this their lives are at greater risk. Douglas Home's Brigadier was so concerned about his possible effect on soldiers' fighting morale that he wrote a report suggesting that Douglas Home should be permanently removed from the front line and found alternative employment. This was not some harsh punishment regime against a selective conscientious objector, but a pragmatic and kindly solution to the problem that he presented. In more recent times, those who resign in protest about some military policy or other are not disciplined for doing so; rather they are allowed to serve their time out, away from front line activities, until they can leave.

The British military is a family community that will find ways to protect itself and to deal with its members who are not fully committed to the community and to its goals. It has innumerable devices available that it can use with its members whose consciences trouble them. They can be posted elsewhere, discharged, given unpaid leave or sent on long courses. Such administrative practices may be a disappointment to those who wish to see gross violations of the human rights of

---

89 Major General Jeremy Reilly DSO, quoted in 'Thoughts on Command in Battle', *British Army Review*, 153 (Winter 2012), pp. 92-100 and p. 97.

90 Tim Collins, 'Who dares put our regiments to the sword?', *The Daily Telegraph*, 8 May 2012, http://www.telegraph.co.uk/news/uknews/defence/9252409/Who-dares-put-our-regiments-to-the-sword.html, accessed 28 September 2013.

91 Christopher Hope, 'David Hart Dyke, the captain of HMS Coventry recalls the horror of his ship sinking in the Falklands War', *The Daily Telegraph*, 18 May 2012, http://www.telegraph.co.uk/news/worldnews/southamerica/falklandislands/9272406/David-Hart-Dyke-the-captain-of-HMS-Coventry-recalls-the-horror-of-his-ship-sinking-in-the-Falklands-War-It-was-black-with-people-on-fire....html, accessed 28 September 2013.

conscientious objectors occurring in the British military. In reality, the practice appears kindly and respectful of individuals' views.

The commonly accepted view is that Britain strongly opposes selective conscious objection by serving military members, but the evidence for this is sparse. Air Commodore Charlton resigned over the bombing of Iraq in the 1920s, but was reinstated and given another job elsewhere. His eventual early compulsory retirement may have had nothing to do with his conscientious protest. Captain Douglas Home was probably right in reasoning that his regiment would have protected him over his conscientious refusal to fight and in doing so avoided controversy. British Army regiments are proud, close knit families that do not want unfavourable publicity.

Douglas Home's undoing was that he publicized his conscientious objection by writing without the Army's permission to a newspaper about his view of what happened. He thus embarrassed his regiment by publicizing the issue and the matter could no longer be dealt with informally within the Army.[92] Official public action had to take place. Clearly, he had stepped beyond the boundary of acceptable behaviour in military eyes, because he had dishonoured the Army, and he was severely punished for his offence.[93]

SAS trooper Ben Griffin was allowed to resign and given a strong testimonial by his commanding officer, probably because he correctly used the military system of his chain of command to quietly make his protest. Two soldiers who made it known that they were unwilling to fight in Iraq, since they were conscientious objectors, were returned home and treated in a manner that allowed an official spokesperson to claim that no conscientious objectors were known. The soldiers were probably quietly discharged. Military officers and civil servants who have crises of conscience are allowed to resign; they are not disciplined because of their behaviour – unless they behave unreasonably.

The cases where courts martial consider conscientious objection are rare and they may not be what they seem. Those cases that do go to court martial seem to be ones where the individual has refused legitimate orders and then pressed their case as a matter of conscientious objection. No military can tolerate such behaviour from its personnel. Courts martial are unwilling, it seems, to acknowledge conscientious objection motives in such cases. Pilot Officer Kenyon at Suez gave confused evidence and it was not clear what his motives were for aborting his attack. Possibly his legal advisors had instructed him to steer clear of matters of conscientious objection in his defence. If he had simply spoken first to his commanding officer, rather than publicly damaged his aircraft and put other's

92 This is also the view of Home's biographer, a retired British Army General. See David Fraser, *Will: Portrait of William Douglas Home* (London: Andre Deutsch, 1995), p. 223.

93 Informed military analysis of the issues at Le Havre at the time concludes that Home was wrong and that he oversimplified the military issues. Ibid., p. 233.

lives in danger, it is inconceivable that he would not have been quickly posted away from flying duties.

Lance Corporal Glenton went abroad absent without leave, returned after two years and initially made no case about his conscientious objection. He then spoke without permission to the media, including publicizing a letter he wrote to the Prime Minister. If he had simply spoken to his commanding officer and not gone absent without leave or publicized his beliefs until he had left the Army, it is likely that he would have avoided court martial. Even then his sentence was reduced from, possibly, over ten years in jail to five months. Leading Aircraftsman Mohisin Khan went absent without leave and refused to discuss his concerns with the RAF – and then took this small matter to the High Court. Leading Seaman Lyons made his case at court martial and appealed, but the decision went against him. His court martial was unwilling to make a judgement about the sincerity of his conscience motives. Trying Lyons for refusing to obey an order to report for pre-deployment training avoided discussion of the legality of the Iraq war at the court martial. Similarly, Kendall-Smith, the RAF doctor who was court martialled for his refusal to deploy to Iraq a third time, was told by the court martial that it thought he wanted to become a martyr. It refused to consider his views about the legality of the Iraq war and instead focussed on his refusal to obey orders to undergo predeployment training.

The cases of Kahn, Lyons and Kendall-Smith, all of whom were medical personnel, raise another issue. It may be that British state and military decision makers, perhaps unconsciously, expect medical personnel, who have promised to help others, to be willing to do so when asked, especially when their role is often not a dangerous one. Support for this argument can be found in the words of Ernest Bevin, the Minister for Labour in 1942. Faced with conscientious objectors reluctant to serve in hospitals he wrote that he

> should like it to be clearly understood that no conscientious objection to serving in hospitals or other similar peaceful occupations can be recognised…I am confident that there is no substantial support in this country for conscientious objection to works of mercy of this kind.[94]

In the both the First and Second World War a Non-Combatant Corps was set up so that conscientious objectors could serve in the Army without bearing arms.[95] In the Second World War some such went into battle as unarmed paramedical personnel and others served as bomb disposal personnel, so there is a history of military service here.[96]

---

94 Cited in Hazel Nicholson, 'A Disputed Identity: Women Conscientious Objectors in Second World War Britain', *Twentieth Century British History*, 18/4 (2007), pp. 409-28 and p. 413.

95 Rae, *Conscience*, p. 192.

96 Peter Brock, '"Excellence in Battle": British Conscientious Objectors as Medical Paratroopers, 1943-1946', *War and Society*, 22/1 (2004), pp. 41-57.

Cases of conscientious objection that reach court martial are very rare. In all the cases cited in this paper there are factors that cast doubt on the claim of conscientious objection. The cases that reach court martial are likely to be ones where individuals are determined to have their day in court, but who, probably, were offered, or could have found, alternative ways of getting out of the situation before the court martial.

Conclusion

British civilians and military personnel who have objected on grounds of conscience to military service are tiny in number. The evidence points to two main conclusions about the way they have been treated. During the two periods of conscription around the First and Second World Wars, Britain tolerated universal conscientious objectors and allowed them to not fight. It also, probably, tolerated selective conscientious objectors as long as they did not offend the military ethos by drawing publicity to themselves and embarrassing their service. Consequently, such cases are unknown. During the periods of all volunteer forces Britain has dealt with selective conscientious objectors through an approach characterized by common sense administrative leniency. The aim has been to remove such people so they cannot raise issues publicly that embarrass the military and government.

The evidence also suggests that the military do not wish to fight alongside someone who may be untrustworthy due to their conscientious objection. Hence they take administrative action to avoid this. The military has shown no interest in deliberating over anyone's conscience. Its attitude has been that an objector will be discharged or moved elsewhere unless military law has been broken in a deliberate manner, in which case, if the matter is serious enough, it will go to court martial. Only when someone has been determined to make a point about their conscientious objection has the issue then risen to public prominence. In these few cases, courts martial have been unwilling to accept conscience as the issue and they have decided the matter on other, straightforward military grounds, such as disobedience of orders. In the 1920s Trenchard had succinctly expressed the British military view about the matter of conscientious objection, when he replied to Air Commodore Charleton: 'An inquiry into what? Your conscience? Certainly not'.[97]

97 Boyle, *Trenchard*, p. 511.

Chapter 9

Selective Conscientious Objection: Philosophical and Conceptual Doubts in Light of Israeli Case Law

Yossi Nehushtan

Introduction

Most states do not tolerate what they perceive as selective conscientious objection. Yet many states which do not tolerate selective conscientious objection do tolerate (to various degrees) what they perceive as non-selective conscientious objection. In this chapter I argue that there are no prevailing justifications for only tolerating non-selective conscientious objections. That is, if a state is willing to tolerate non-selective conscientious objections, it must also tolerate selective conscientious objections to a similar degree (all other things being equal). Specifically, I claim that there is no such thing as 'selective' conscientious objection, or rather that conceptually, all conscientious objection, even the type mistakenly labelled 'absolute' conscientious objection, is in fact selective conscientious objection. Moreover, even if the conceptual distinction between selective conscientious objection and non-selective (or absolute) conscientious objection is sustainable, I argue that this distinction has no significant moral or practical implications and that ultimately all kinds of conscientious objection should be treated equally (all other things being equal).

In the second section of this chapter I lay out a few definitions of the central concepts (conscientious objection, selective conscientious objection and civil disobedience) in order to avoid misconceptions. In the third section, I argue that the main rationale for granting conscientious exemptions involves acknowledgement that the conscientious objector does not have real choice when he acts upon his conscience. In the fourth section, I explain why selective conscientious objection is merely a form of conscientious objection and therefore should be treated the same. And in the fifth section, I describe and scrutinize recent Israel Supreme Court case law that dealt with the issue of selective conscientious objection.

I wish to note at the outset that the scope of this chapter excludes an exhaustive discussion of the complex issues in selective conscientious objection in general, selective conscientious objection to army enlistment, or even the implications of Israeli case law. This chapter merely offers a partial introduction to the general issue and a few critiques of recent thought-provoking Israeli case law. Also, this chapter does not differentiate between conscripts and volunteers. Since the Israel Defence Force (IDF) is a conscript army, all cases of conscientious objection within Israeli case law involve conscripts. Yet, the theoretical arguments about selective conscientious objection alongside the conceptual analysis which are offered in this chapter apply to conscripts and volunteers alike.

Conscientious Objection, Selective Conscientious Objection and Civil Disobedience

Conscientious objection is a private act whose only purpose is to distance the objector from acting according to a legal demand. With regard to military service, a conscientious objector refuses to join the army for 'conscientious reasons', that is, because they think that they are, regardless of the circumstances, morally prohibited from enlisting into the army. This objection is often perceived as absolute because it is based on an absolute objection to the use of lethal measures or violence.

Selective conscientious objection is also a private act, whose only purpose is to distance the objector from acting according to a legal demand. With regard to military service, the selective conscientious objector does not necessarily oppose to enlisting in the army altogether and under all circumstances, but rather opposes the particular war, military operation or policy he or she is being asked to engage in. As long as the war, operation or policy is sustained, the selective conscientious objector refuses either to take part in the fighting or operation, or to join the army altogether. Such a refusal may be based on either the aim of the war or the policy or the manner in which the war or policy are carried out. This objection is perceived as selective because it results from the objector's assessment of the morality of certain wars, operations or policies, as opposed to his assessment of the overall morality of military service as such.

Civil disobedience is a public, political act, whose purpose is to bring about a change in the law or in existing policies. Whereas conscientious objection is a breach of a law that the agent is morally prohibited from obeying, civil disobedience is politically motivated. According to Joseph Raz, civil disobedience is essentially a public action, while conscientious objection is essentially a private action. Raz argues that the case for a right to conscientious objection seems much stronger.[1] This approach is similar to that of many scholars and governments, who are willing to tolerate conscientious objection to some extent, but not civil disobedience. However, Raz also mentions that these categories are partially overlapping.[2] Conscientious objection and civil disobedience can interact in two kinds of ways. First, a specific act can be both conscientious objection and civil disobedience, for example, when the conscientious objector also wishes that their

1 J. Raz, *The Authority of Law* (Clarendon Press, Oxford, 1979), p. 276.

2 Raz, *The Authority of Law*, pp. 264 and 276.

refusal to serve in army would promote a change in the law. Second, there can be two or more different acts involved: a private refusal to serve, complemented by additional public acts of a political nature that the objector himself takes on, which might be unlawful but do not necessarily have to be.

Conscientious Objection's Grounding and Rationale

According to the common view, conscientious objection is grounded in autonomy or in freedom of conscience. Put differently, conscientious objection is tolerated out of respect for the objector's autonomy or his freedom of conscience.

Emphasizing freedom of conscience or autonomy as a central concept within the issue of conscientious objection implies that the conscientious objector should have a choice or that he or she does in fact have a choice. However, it is a truism that one cannot choose one's deepest beliefs. One also does not always have a meaningful choice about whether to act upon one's beliefs. Therefore, with regard to granting conscientious objection, we should apply the concept of autonomy with caution, as granting conscientious exemption does not reflect respect for the conscientious objector's right to choose but rather acknowledges his lack of real ability to refrain from acting upon his conscience, i.e. his lack of real choice. This situation is demonstrated in the case of having to choose between (a) not acting upon one's conscience and (b) suffering a meaningful sanction should one act upon it (e.g., a Muslim woman who feels compelled by her religious conscience to wear a veil and faces a criminal law that utterly prohibits the wearing of a veil). This is not necessarily the case, however, when faced with a choice between (a) not acting upon one's conscience and consequently enjoying various types of benefits and (b) acting upon it and thereby losing these benefits (e.g., a Muslim woman whose her religious conscience compels her to wear a veil, facing a law that utterly prohibits wearing a veil in public universities but not in numerous private ones). Only with regard to the second case can we identify real choice and, as a result, acknowledge the importance of autonomy as a guiding principle within conscientious objection.

Generally speaking, freedom of conscience does not protect the right to *choose* to do X (or not to do X), as much as it protects the right to do X (or not to do X). The 'will theory' or the 'choice theory' of rights may be sound, in the sense that one usually has a choice regarding whether or not to exercise any right or liberty that one possesses. However, a person compelled by his conscience to do X or to avoid doing X is simply seeking a legal right to act upon his conscience, which may mean possessing a legal right to do what he would do anyway, because he does not have a real choice in this matter. Therefore, having a legal right in these cases does not necessarily entail a choice regarding whether to exercise it.

Brian Leiter asks why protecting liberty of conscience promotes human well-being. His answer is that, 'Many of the arguments trade, at bottom, on a simple idea: namely, that *being able to choose what to believe and how to live* ... makes for a better life. Being told *what you must believe* and *how you must live*, conversely, make lives worse.'[3] I suggest, however, that being told what to believe and how to behave is not the only alternative to being able to choose what to believe and how to live. A third option exists, according to which protecting a person's right to have and act upon a certain conscience promotes the person's well-being because he cannot really control his conscience, i.e. select what to believe in and, at times, how to live.

Before further discussing the argument about 'lack of choice' (yet without exploring in depth arguments from determinism) a distinction should be drawn between possessing a certain conscience and acting upon it. With regard to having a certain type of conscience, the 'lack of choice' argument applies strongly. A person cannot choose their conscience. It is hard, perhaps even impossible, to comprehend why and how a person ends up having a certain conscience and not another. An individual cannot simply will to change their conscience, beliefs and views. They may be convinced of the rightness or wrongness of certain arguments and, accordingly, may change their conscience, beliefs or views, but they cannot select to make such a change in the same way as they choose what to have for lunch or where to spend their next holiday. Even in the case of a changed conscience, beliefs or views following persuasion by certain arguments, an individual cannot control their ability to be persuaded by certain arguments and not by others, and therefore the 'lack of choice' argument remains intact.

However, with regard to acting upon one's conscience, the 'lack of choice' argument applies in a weaker (yet still meaningful) way. It may be argued that one does have a choice regarding whether or not to act upon one's conscience. Even when the only options are either to act against one's conscience and live or to act upon one's conscience and consequently suffer a horrible death, the choice still exists. This is, of course, an overly narrow and, therefore, unsatisfactory perception of what possessing a choice means. A broader, more desirable notion of possessing choice would hold that normally, a conscientious objector does not really have a choice about whether to act upon their conscience, since the result of not acting upon it would bring devastating harm to their personhood.

Thus, one way to justify granting conscientious exemptions is to turn to the principle of autonomy that states that people should be able to make their own choices and have an adequate range of good options from which to choose. A different, presumably better way to justify granting conscientious exemptions is to acknowledge that in matters of conscience, people do not have a real choice. This lack of choice results from the facts that one cannot select one's conscience and that acting contrary to one's conscience results in severe harm to one's personhood.

However, the distinction suggested above between the autonomy justification and the 'no real choice' justification for tolerating conscientious objections may not be so absolute. Having no real choice regarding whether or not to act in a certain way may well affect one's autonomy if the choice in this particular matter

---

3 B. Leiter, 'Why Tolerate Religion?', *Constitutional Commentary*, 25/1 (2008), p. 25.

is exceptionally important for the individual. If one is coerced to act contrary to one's conscience, since acting otherwise would lead one to suffer meaningful or severe sanctions, one's autonomy is invaded.

If one must act against one's conscience to enjoy a certain benefit or option, and that option is exceptionally valuable to one, then one's autonomy is invaded. For example, if a person's religion compels him to wear a yarmulke, and if wearing yarmulkes is prohibited in all schools and universities, yet being a teacher is the person's ultimate desire as a means to fulfil his aspirations, then the de-facto lack of an adequate range of options invades his autonomy. Having an adequate range of options does not only mean having a sufficient number of options. In the example above, even though the individual still has other career options that can be perceived as valuable, these options are of no comfort if all he really wants to do is to teach and no other career can provide him with even half the satisfaction and fulfilment. A person has a real choice and can be seen as autonomous if they have an adequate range of valuable options in a subject that matters to them, and if these options are valuable from that person's perspective (and perhaps also objectively). Thus, the fact that one has no real choice with respect to one's conscientious beliefs may raise issues regarding one's autonomy after all.

All the above is true regarding selective conscientious objection and non-selective conscientious objection alike. The applicability of the 'no real choice' argument to selective conscientious objection is relevant to the question of whether selective conscientious objection in fact exists, which will be discussed below, and to the discussion about Israeli case law that follows.

Do Selective Conscientious Objections Exist?

Until now I have assumed that selective conscientious objection exists. At this point, however, I wish to maintain that the concept of selective conscientious objection, or the assumption that there are meaningful differences between selective conscientious objection and non-selective conscientious objection, is unsustainable.

Back in 1979, Raz implied, without elaborating on the point, that a 'selective' conscientious objection is merely a kind of conscientious objection, emphasizing that it is first and foremost conscientious objection and therefore (other things being equal) it should be treated like any other conscientious objection.[4] Since then, other scholars have expressed similar views, especially but not exclusively within the narrower context of unjust wars.[5] Yet, most courts and administrative authorities

4 Raz, *The Authority of Law*, p. 263.

5 P. Robinson, 'Integrity and Selective Conscientious Objection', *Journal of Military Ethics*, 8/1 (2009) p. 34; J. Wolfendale, 'Professional Integrity and Disobedience in the Military', *Journal of Military Ethics*, 8/2 (2009), p. 127; G. Clifford, 'Legalizing Selective Conscientious Objection', *Public Reason*, 3/1 (2011), p. 22.

in most countries still do not tolerate what they perceive as selective conscientious objection, while they are willing to tolerate what they perceive as non-selective conscientious objection. This is troubling because the case against the distinction between selective and non-selective conscientious objection is quite strong and because it creates a disturbing gap between the views of expert scholars in the field on the one hand and administrative policy and court decisions on the other.

From a conceptual perspective, there is no selective, non-selective (or absolute) conscientious objection, but just conscientious objection. This is the case for two main reasons.

First, the distinction between selective and absolute objection can only be made by someone other than the conscientious objector themself. State agents normally undertake this differentiation. The objector themself cannot really choose between opposing army enlistment in general, opposing all wars and opposing a specific war. One's conscience has its own will and its own boundaries, and they may be broad or narrow, conditional or unconditional. This is hardly a matter of choice. Classifying an objection as 'selective' implies that the objector 'selects' what they object to; yet the true conscientious objector does not and in fact cannot control their conscience – and in turn cannot select the scope of their conscientious objection.

Second, almost all objections can be perceived as 'selective'. If by 'selective' we refer, wrongly, to the scope of the objection, then selective objection is in fact a narrow objection in comparison to broader objection. Objection to a specific war is more selective or narrower than objection to all wars (which is mistakenly known as 'absolute' objection). However, objection to all wars is more selective or narrower than objection to join an army. Moreover, the objection to join an army is more selective or narrower than an objection to participate in any activity that benefits the army (such as, for example, working in a factory that provides food, clothes or military equipment to the army). Objection to participate in any activity that benefits the army is also selective in respect to an anarchist's political objection to treat any law as authoritative – or to an objection to obey any secular law as opposed to religious law; and so forth. The only true 'absolute' or non-selective objection is an objection to obey all laws. All other objections are in fact selective to various degrees. The fact the some objections are 'more selective' or narrower than other objections bears no moral implications. It is none other than a rather uninteresting fact.

If by 'selective' we do not refer to the scope of the objection but rather to the reason for the objection, and by that we are in fact equating 'selective' with 'specific' or 'conditional', then yet again, almost all objections can be perceived as 'selective'. A conscientious objector can refuse to enlist in the army altogether. They can do so for various conscientious reasons that may be selective or specific to varying degrees. They can refuse to join the army because the army was responsible for their parents' death but never agreed to be held accountable; because the army is currently involved in an unjust war; because the army constantly assists the government in persecuting political minorities; because he or she is a pacifist or an anarchist. All these conscientious reasons, apart from the latter, are selective or specific. They all lead to an absolute refusal to join the army. This example shows that from the conceptual point of view, there is little point in distinguishing between kinds of objection and classifying them as selective and non-selective. Once the conceptual distinction fails, it becomes hard to comprehend how the selective, specific or conditional nature of a conscientious objection can bear any moral importance or moral implications.

One common argument against selective conscientious objection is that it is in fact identical to civil disobedience or 'political objection', or at least is closer to civil disobedience than to a 'pure' conscientious objection. As I will show below through a discussion of Israeli case law, Judges tend to equate selective conscientious objection with civil disobedience, or to claim that the selective conscientious objection is in fact 'political' or 'ideological' rather than conscientious, thereby justifying its dismissal.

Much can be said about the above argument against selective conscientious objection. I will only discuss this issue in a nutshell. The argument that selective conscientious objection is in fact 'political' objection can have two meanings. It may mean that even though the objection is private in nature and even though the objector does not aim to change the law, the objection is grounded in 'political' rather in 'conscientious' reasons. That is, the objector disobeys the law because they think that the law is misguided yet do not think they are under a moral-conscientious duty to disobey the law. I will not discuss this case here, but suffice it to say the judges often do not defend such a conclusion. They merely attach the term 'political' to the objection in order to not tolerate it for other reasons. The argument that selective conscientious objection is in fact 'political' objection may also mean that the objector in fact wishes that their objection (perhaps together with further acts) would be part of a political campaign to change the law.

In the second case the distinction between civil disobedience and conscientious objection does exist and may be helpful, but only to a certain extent. The differentiation collapses when we face acts of public disobedience whose purpose is to bring about a change in the law or in existing policies because the disobedient person believes they are under a moral, conscientious duty to bring about a change in the law or in existing policies. Even if we classify selective conscientious objection as civil disobedience, it does not necessarily follow that it should not be tolerated as such. If a selective conscientious objector or a civilly disobedient person commits or refrains from an act because they think they are under a moral duty to do so, i.e. because their conscience compels them to do so, then the harm caused to such an objector by forcing them to obey the law is by no means different than the harm caused to an 'absolute' conscientious objector. All rationales for granting conscientious exemption (autonomy, freedom of conscience or acknowledging the objector's lack of choice) apply here in the same manner. Accordingly, all the reasons for not granting conscientious exemptions also apply here in a similar manner.

Courts and policy makers may agree that from a conceptual point of view, all conscientious objections are selective. They may also agree that selective conscientious objection is just a kind of conscientious objection. However, they would and indeed do argue that there are still meaningful differences between total, pacifist refusal to enlist to the army and conditional refusal to do so. These alleged differences normally rely on prudential reasons. These reasons will be addressed below as part of the discussion of the Zonshein case.

The Unfortunate Israeli Case

The following discussion will focus exclusively on Israeli cases of selective conscientious objection regarding the armed forces. Israeli case law and policy have produced brilliant examples of glorious mistakes regarding the issue of selective conscientious objection.

The Israel Defence Forces (IDF) are a conscript army. Men and women are recruited at the age of 18, but whereas men serve three years in the IDF women only serve two. Men may be recruited to combat forces, even against their will. Women may volunteer for several combat positions and will then serve for three years. After completing their compulsory service, men and women may be called up to reserve military service in which they are called for active duty, which may last from several days to several weeks every year.

In Israel, since the late 1980s, the public and legal discussion of selective conscientious objection has focused on three kinds of cases. In the first case, the selective conscientious objection results from the on-going occupation of the West Bank and the Gaza Strip. In this case, some objectors do join the army and only refuse to participate in acts that are committed in the occupied territories or directly sustain the occupation. Other objectors refuse to join the army as long as the occupation is sustained.

In the second case, the selective conscientious objection resulted from the government decision to withdraw from the Gaza Strip in 2005, and currently from sporadic government decisions to evacuate illegal settlements. Most of these objectors do join the army and are interested in continuing their military service, even in combat forces, but only refuse to take part in evacuating Jewish settlers from their homes in the occupied territories. The first two kinds of objection, the first a left-wing selective objection and the second a right-wing one, are indeed widely considered selective conscientious objection.

The third case, however, is quite absent from the debate about selective objection and is not perceived as such. In the third case, the selective objection results from a religious point of view. These religious selective conscientious objectors are willing to join the army but demand that the army accommodate their religious beliefs, which include demands not to be exposed to women singing in public; to women they consider to be dressed in an 'indecent' manner; to female instructors and other occurrences they consider offensive. Other religious selective objectors will only join the army as part of designated military units (*Yeshivot Hesder*) in which they serve 18 months rather than the obligatory 36 months military service and enjoy numerous social and cultural privileges. This last type of conscientious objection has never been discussed in Israeli courts and will not be covered here.

I will only briefly review Israeli case law of selective conscientious objection, without discussing any decisions from over a decade ago.[6] I will limit the discussion to the current underlying approach of the Israeli Supreme Court towards selective objection and will scrutinize it in light of the previous discussion about the nature of selective conscientious objection. In the last decade, the Israeli Supreme Court has delivered three landmark decisions regarding selective conscientious objection: the Zonshein case in 2002,[7] the Milo case in 2004[8] and the Shvili case in 2010.[9] I will discuss these cases in chronological order.

The Zonshein Case: Establishing the False Distinction between Selective and Non-Selective Conscientious Objection

The Zonshein case dealt with the refusal of Lieutenant David Zonshein to serve in the occupied territories.[10] The IDF decided to not tolerate what it perceived as selective conscientious objection, and the Israeli Supreme Court upheld this policy. The decision was based on the following main reasons:[11] first, selective objection is more common than non-selective objection and has the tendency to spread, and

6 For those decisions see B. Medina 'Political Disobedience in the IDF: The Scope of the Legal Rights of Soldiers to be Excused From Taking Part in Military Activities in the Occupied Territories', *Israel Law Review*, 36/3 (2002), pp. 73 and 75-6; A. Paz-Fuchs and M. Sfard, 'The Fallacies of Objections to Selective Conscientious Objections', *Israel Law Review*, 36/3 (2002), pp. 111 and 113-15.

7 H.C 7622/02 *Zonshein v. Judge Advocate General* 57(1) P.D. 726 (2002). For the English translation see: 'Zonshein v. Judge Advocate General', *Israel Law Review*, 36 (2002), p. 1.

8 H.C 2383/04 *Milo v. Ministry of Defence* 59(1) P.D 116 (2004). The Milo decision has not been published in English.

9 H.C 5587/09 *Shvili v. Ministry of Defence* (28 July 2010). Petition for further hearing was denied in H.C 5977/10 *Shvili v. Ministry of Defence* (31 August 2010). These decisions have not been published in English.

10 For more details see Paz-Fuchs and Sfard, 'Fallacies', pp. 116-17: In January 2002, 50 conscientious objectors published a petition declaring their refusal to fight beyond the 1967 borders. The objectors declared their willingness to continue to serve in the Israeli army in any mission that serves its defence, but at the same time made clear that they do not see 'missions of occupation and oppression' as serving that purpose. The 'exclusivity' of the group, consisting of individuals who served in the past as combat officers and soldiers, highlighted the selective nature of their objection. They protested not against the existence of an army as such, but rather against the role Israel's army plays in the occupied territories. Within several months, the group's number rose to over 500, an unprecedented number of conscientious objectors in Israel. By May 2002, over a hundred group members were sentenced to prison terms by their commanding officers.

11 See also Medina, 'Political Disobedience', p. 90. For further reasons and for detailed and convincing critique see Paz-Fuchs and Sfard, 'Fallacies'.

thus it threatens society as a whole or the state itself; second, tolerating selective conscientious objection would jeopardize social and Army unity and society's sense of solidarity; third, an evidential problem makes it difficult to determine who is a selective conscientious objector and who is 'just' a political objector and to assess the magnitude of people's moral or conscientious objection to a specific assignment. I shall discuss these reasons in turn.[12]

Do Numbers Matter?

The number of those who seek conscientious exemption is indeed a relevant factor. Conscientious exemptions may be granted up to a certain quota, since exceeding that quota will diminish the purpose of the general rule itself. Assume that the law exempts religious believers from legal compulsory enlistment in the army. It may be the case that national security, which is the rationale of the general rule, would allow no more than X number of exemptions from the rule at any given time or for any given set of circumstances. It may also be the case that the number of sincere religious conscientious objectors exceeds X. The state now faces three options: to grant exemptions to all the sincere objectors, thus jeopardizing national security; to grant exemptions to none and thus harm conscientious objectors more than the public interest demands; and to grant exemptions only to certain sincere objectors, i.e. to X number of them. I argue that the government should generally exempt 'selective' objectors as long as their number does not jeopardize the public interest. The mere fact (if this is indeed a fact) that there are more selective conscientious objectors than non-selective ones, or the alleged fact that selective objection has a tendency to spread, does not in and of itself justify a general rule against granting conscientious exemptions to all selective conscientious objectors as such. Medina refers to this issue and posits that:

> The deep controversy within Israeli society over the government's policies regarding the Occupied Territories and the fight against terror, and thus the intense public debate regarding the IDF's activities in these fields, creates a significant potential, that meets even the high standard of 'near certainty' of widespread objections to serve, once 'political' objections are tolerated.[13]

In contrast, Paz-Fuchs and Sfard take the opposite view, claiming that 'the fear of mass disobedience that will spread like a plague seems to be more in the minds of politicians and lawyers, and stands contrary to research in the field of social psychology'.[14]

---

12 This is not an exhaustive list of the pragmatic justifications to not tolerating selective conscientious objection. For further (alleged) justifications and for convincing critique see: Clifford, 'Legalizing', pp. 33-5; Paz-Fuchs and Sfard, 'Fallacies'.

13 Medina, 'Political Disobedience', p. 99.

14 Paz-Fuchs and Sfard, 'Fallacies', p. 127.

Either way, the state cannot use the real or the alleged tendency of selective conscientious objection to spread as a reason to not grant conscientious exemptions to all selective conscientious objectors. If the state does recognize selective objectors to be selective *conscientious* objectors (as at times it must), it should exempt 'selective' objectors as long as their number does not jeopardize the public interest. If eventually the numbers grow, the state would have to grant exemptions only to some of the objectors. The state must find the solution that would maximize the protection given to freedom of conscience without giving up the rationales of the justified general legal rule, namely it must grant exemptions only to some objectors. The selection process can result from a ballot that, however arbitrary, reflects equality of opportunities. Tolerating the maximum justifiable freedom of conscience should be the governing principle. Freedom of conscience is no different from other goods that can be distributed only to some who are relevantly equal to others and have equal entitlements to these goods. If a local government body can finance shelters that can accommodate 200 homeless people, but 300 homeless reside in its jurisdiction, there is not much sense in accommodating none of them in the name of equality. If the distribution process itself is fair, as in the case of a ballot, the claim to equality loses much of its strength and cannot justify denying the goods to everyone. Regarding conscientious exemptions, our main concern should be tolerating the objectors' conscience as much as possible, rather than as equally as possible, when both aims cannot be achieved simultaneously.[15]

The Solidarity Argument

As Paz-Fuchs and Sfard pointed out, Justice Aharon Barak expressed concern that 'the people's army will transform into an army of peoples comprised of various units, to each of which it would be conscientiously acceptable to act in certain areas, whereas it would be conscientiously unacceptable to act in others.'[16] This is a very persuasive argument, but not against granting exemptions to all selective conscientious objectors. Granting exemptions to secular-humanist individuals who are sincere selective conscientious objectors would not lead to forming distinct units within the army which, as such, would object to performing specific military tasks. In Israel, left-wing humanist, mostly secular selective conscientious objectors are small in numbers (a few hundred at the most) and are assigned as individuals in all army units. The solidarity and unity argument is a persuasive argument against enlisting ultra-Orthodox Jews to designated units within the army. Enlisting ultra-Orthodox Jews into designated units within the army

15 This is a slightly different argument than the common liberal anti-egalitarian one or the 'priority v. equality' one. For this type of arguments see for example: H. Frankfurt, 'Equality as a Moral Idea', *Ethics*, 98/1 (1987), p. 34; H. Frankfurt, 'Equality and Respect', *Social Research*, 64/1 (1997), p. 6; D. Parfit, 'Equality and Priority', *Ratio*, 10/3 (1997), p. 202.

16 Paz-Fuchs and Sfard, 'Fallacies', p. 125.

(*Yeshivot Hesder*), a long-standing practice for religious soldiers and a growing practice regarding ultra-Orthodox soldiers, does in fact raise such concerns, which will not be elaborated here.

The Evidential Problem

The last argument against tolerating selective conscientious objection focuses on the evidential problem. According to this claim, it is hard to decide who is a selective conscientious objector (who may be tolerated) and who is 'merely' a political objector (who should never be tolerated).[17] Indeed, deciding who is a selective conscientious objector (or simply a conscientious objector) and who is 'merely' a political objector is difficult. Granted, no mechanism is likely to avoid unequal decisions. However, this is still a lame excuse for not granting conscientious exemptions to selective conscientious objectors. First, the burden to prove honesty lies on the objector's shoulders. It is not the state that has the prove dishonesty. Second, proving honesty is a problematic issue regarding 'absolute' objections as well. Third, these kinds of decisions are not so different from other decisions a court (and occasionally an executive) should and does make.

The Milo Case: More of the Same – and a Bit Worse

The Milo case dealt with Leora Milo, a female who refused compulsory enlistment into the army because of the on-going Israeli occupation of the West Bank and the Gaza Strip. Milo specifically stated that she was not a pacifist and she would have enlisted into the army if it did not sustain the occupation of Palestinian territories. The 'Conscience Committee' thus decided that Milo's objection does not rely on her conscience but rather on her political views. This case gave rise to numerous questions that will not be discussed here (among these, for example, was whether a difference exists under Israeli law between men and women who are conscientious objectors). I will focus exclusively on the question that deals with the difference between selective and non-selective conscientious objection.

17 See also Clifford, 'Legalizing', p. 23: 'proponents of selective conscientious objection often have failed to move even those policymakers who seem sympathetic to selective conscientious objection to act, for the latter have contended that creating a selective conscientious objection option entails an administrative nightmare'. But compare this to Paz-Fuchs and Sfard, 'Fallacies', p. 139: 'It is indeed recognized that an administration true to realizing its mandate has to be wary of the fact that a system that inspects the beliefs of individuals is prone to some evils such as abuse, on the one hand, and public intrusion into private affairs, on the other. But, once again, the same would apply to the onerous duty that is the part of administrative conscience committees that are charged to inspect the nature of a claimant's putative pacifism.'

Justice Ayala Procaccia acknowledged that no strict distinction exists between conscientious objection and civil disobedience. She also acknowledged the existence of what may be called 'hybrid private objection', in which the objector refuses to perform an act both because performing the act contradicts his conscience and because he wishes that his refusal, usually together with that of others, would promote a change in the law or in certain policies.[18] Justice Procaccia also distinguished between selective and non-selective conscientious objection. Her definitions of both types of objection are fairly close to those given above (distinguishing between unconditional and conditional refusal to join the army). As such, they repeat the fallacy of differentiating between things that are not really distinct. Yet problematic as it may be, the conceptual fallacy is not the main issue here. The real problem results from Justice Procaccia's following assumption. According to Justice Procaccia, non-selective objection to military service (which relies on pacifist views) is almost always a pure conscientious objection, whereas selective objection to military service tends to be 'hybrid objection', in which the objector refuses army enlistment for both conscientious and political-ideological reasons. Moreover, Justice Procaccia claims that selective objection primarily comprises civil disobedience, though it may also involve 'conscientious elements'.[19] This reasoning suffers from several inaccuracies.

First, Justice Procaccia does not compare non-selective objection with selective objection. She compares non-selective objection *that relies on pacifism* with selective objection. This comparison is meaningless. To say that non-selective objection that relies on pacifism is almost always pure conscientious objection is circular logic, like saying that pure conscientious objection is pure conscientious objection. What in fact should be compared are non-selective and selective objections.

Second, if we do compare non-selective and selective objections, assuming that 'selective' in this case refers to the scope of the objection, then the argument above is wrong on both accounts. Non-selective objection, that is to say an unconditional objection to enlist to the army, does not have to result from conscientious objection. It can result from reasons that have to do with comfort, fear, egotism and so on. Selective objection, that is to say a conditional objection to enlist, to participate in specific operations or to obey specific orders, may well be grounded in conscience. Nothing in the 'selective' nature of an objection leads to a conclusion that the act of objecting comprises primarily civil disobedience and at best may also involve 'conscientious elements'. No logical relation exists between the selective nature of an objection and the type of reason that grounds the objection. Justice Procaccia assumes that the selective nature of an objection, its conditionality, and the particular objection to a specific policy all indicate that the objection is political or ideological rather than conscientious. However, this reasoning is flawed. One can think of numerous examples of selective, conditional

18 *Milo*, para 9 of Justice Procaccia's decision.

19 *Milo*, paras 10 and 14 to Justice Procaccia's decision.

objections that refer to specific policy, which would still be considered pure conscientious objections. Take, for example, the refusal of a religious Jew to enlist in the army because the army is unwilling to provide Kosher food to its soldiers; the refusal of a Jewish soldier to evacuate Israeli settlements from the occupied territories because he believes it contradicts a divine order; the refusal of a racist Jew to enlist in the army because she is not willing to obey orders that are given by non-Jewish soldiers; the refusal of an animal rights activist to enlist in the army because army uniforms or equipment are made of leather; or the refusal of a soldier to take part in a specific operation because the soldiers were ordered to 'take no prisoners'. Regardless of the moral soundness or the legitimacy of these objections, arguing that they do not comprise conscientious objections merely because they are selective or conditional would make no sense.

In the Milo case, classifying Milo's objection as 'selective' provided sufficient grounds for the court to deny the petition. Justice Procaccia was willing to acknowledge that Milo's objection was not merely political or ideological but also conscientious. However, the mere fact that Milo did not object to serving in the army but rather to serving in an army that sustains the occupation of Palestinian territories doomed her petition. Justice Procaccia relied heavily on Justice Barak's judgment in the Zonshein case (delivered unanimously by Justice Barak, Justice Dorit Beinish and Justice Procaccia). The Zonshein decision was severely criticized by quite a few scholars in the 2004 Issue of Israel Law Review. However, since the Milo decision was also published that same year, none of the scholars who published these articles in the Israel Law Review referred to the Milo decision. More importantly, Justice Procaccia did not and could not address the powerful criticism that Israeli scholars aimed towards the faulty distinction between selective and non-selective objections and towards the misguided moral implications that result from it. Therefore, the Milo decision reinforced the Zonshein decision and reasoning without refuting its critics' arguments. Accordingly, the Milo decision is exposed to the same criticism mentioned above regarding the Zonshein case, and it is mistaken for the same reasons. Moreover, the Milo case is even more erroneous than the Zonshein case, precisely because Leora Milo is a female. There is a strong reason to believe that most if not all of Justice Barak's reasoning in the Zonshein case is either inapplicable or much less applicable to female selective conscientious objectors. This is the case, briefly, because a tiny percentage of soldiers in the combat forces are women and because the majority of women in Israel do not enlist into the army. Consequently, one needs a developed imagination to argue that the tendency of selective objection to spread among women might compromise national security.

The Shvili Case: The New Kid on the Block – Institutional Selective Objection

The Shvili case is quite extraordinary. Reuven Shvili was an ultra-Orthodox Jew and a Yeshiva student and was therefore legally exempt from army service at age 18. The law at that time enabled the authorities instead of permanently exempting ultra-Orthodox Yeshiva students from enlistment, to merely postpone their recruitment as long as they remain Yeshiva students. Following a few years of postponing the recruitment, Yeshiva students would receive a permanent exemption from enlistment according to various criteria. Yet at age 21, Shvili quit his Yeshiva studies, gradually embraced a non-religious way of life and applied for an exemption from enlistment for secular-conscientious reasons. His application was denied twice. At first, it was denied because the 'Conscience Committee' decided that Shvili's objection results from reasons of comfort and not of conscience. After a further hearing, the 'Conscience Committee' changed its reasoning and decided that Shvili is indeed a conscientious objector, but not a pacifist, and therefore is not entitled to an exemption.[20]

Shvili's views, which were not disputed, were complex but not necessarily incoherent.[21] He never declared that he was an 'absolute pacifist', but he did declare that he preferred to avoid violent situations and that violence deters or intimidated him. Shvili added that he would not mind, in principle, joining the police forces, as long as he would only fulfil 'desk jobs' that do not relate even remotely to the IDF. Shvili justified his principled objection to join the army by arguing that the army is inherently violent. He also refused to enlist into the army in the capacity of 'non-violent' assignments only, but he was willing to commit to perform 'national/civil service' instead of enlisting.

Shvili's selective conscientious objection is unique and was never discussed in Israeli case law. In Israel, selective conscientious objectors normally base their objection on the immorality, in their view, of specific IDF policies and operations. Thus, 'left-wing' objectors refuse to enlist into an organization that sustains the occupation, to serve in the occupied territories or to take part in unjust wars or unjust military operations that may harm Palestinian civilians. 'Right-wing' objectors, on the other hand, mainly refuse to take part in evacuating Jewish settlers. Religious objectors refuse to listen to women singing in official ceremonies or in cultural events, to be exposed to women in 'indecent' outfits or to be taught by women.

Shvili's objection is different: it is non-conditional and does not result from specific IDF policies. It is indeed a 'selective' objection in the sense that Shvili differentiates between types of state organs that apply violent measures to fulfil their tasks. Shvili was willing to be drafted in to the police forces and to fulfil 'desk-jobs', even though the police is also a state organization that applies violent measures in order to achieve its tasks. He was, however, unwilling to join the army at all or to take upon himself police posts or responsibilities that are connected to the IDF.

Therefore, Shvili's objection is 'institutionally selective' rather than 'ideologically selective'. Yet, with regard specifically to the IDF, his objection cannot be considered selective. Does the difference between 'institutional conscientious selective objection' and 'ideological-conscientious selective objection' bear any moral implications? There are three possible answers to this question.

---

20 *Shvili*, paras 1-3 to Justice Procaccia's decision.

21 *Shvili*, para 3 to Justice Procaccia's decision.

The first answer was elaborated throughout this chapter. I have argued that the distinction between selective and non-selective conscientious objection is conceptually impossible and yields absolutely no moral significance. All that matters is that the objection is conscientious and sincere. Selective conscientious objection, if it exists, is merely a kind of conscientious objection and should be treated identically. The same logic applies to 'institutional conscientious selective objection' and 'ideological-conscientious selective objection' alike.

The second answer was given by Justice Procaccia in the Shvili case. Justice Procaccia maintains the erroneous distinction between selective and non-selective conscientious objection. She repeats the reasoning from the Milo case and the Zonshein case and decides that only pacifist conscientious objectors should be exempt from enlistment in the army. 'Selective' conscientious objectors, however, should not be exempt, even when they truly are conscientious objectors.[22] Justice Procaccia sees no difference between 'institutional selective objection' and 'ideological-conscientious selective objection'. Both are selective objections, and as such should not be tolerated.

The third answer was given by Justice Esther Chayut in the Shvili case. Justice Chayut also maintains the flawed distinction between selective and non-selective conscientious objection, but, unlike Justice Procaccia, does accord moral significance to the sub-distinction between 'institutional conscientious selective objection' and 'ideological-conscientious selective objection'. Justice Chayut rightly notes that with regard to the IDF, Shvili's objection is not selective but absolute, and that generally speaking, Shvili can be classified as contingent pacifist.[23] Justice Chayut also adds that part of the reasoning in the Zonshein case and in the Milo case for differentiating between selective and non-selective objection (reasoning that Justice Procaccia adopted in full in the Shvili case) does not apply at all in this case or applies only marginally, since Shvili's objection to enlist in the army is absolute and institutionally-based rather than 'specific ideology'-based. This is no doubt true. However, Justice Chayut was not willing to take this reasoning a step further and accept the petition. Instead, she decided that even though there are meaningful moral implications to the distinction between 'institutional conscientious selective objection' and 'ideological-conscientious selective objection', the balance of reasons (as laid out in the Zonshein and Milo cases) still leads to a conclusion that selective objection should not be tolerated.[24]

In the Shvili case, the Israeli Supreme Court did not confront at all the academic criticism following the Zonshein case and simply repeated and reaffirmed the Zonshein reasoning. Initially, it may appear that the Shvili case, much like the Milo case, merely offers more of the same mistakes. However, winds of change can be found in the Shvili case after all.

---

22 *Shvili*, paras 12-14 and 17 to Justice Procaccia's decision.

23 *Shvili*, para 4 to Justice Cahyut's decision.

24 *Shvili*, para 5 to Justice Cahyut's decision.

First, in her concluding remarks, Justice Procaccia stressed that although the petition was denied (meaning that that Shvili should be drafted and treated like any other soldier), she recommended that the authorities accommodated Shvili's conscience by assigning him to posts that involved no direct contact with violent acts. Perhaps this is not what Shvili was hoping for, but it is more than Zonshein and Milo achieved. Zonshein, an army reserve officer who as such serves in the army for about 30 days a year, only requested to not be assigned to military duty in the occupied territories. He was denied completely. Milo refused to enlist into the army as long as the occupation continued. Not only was she denied, but she was rebuked by Justice Procaccia, who said that 'the petitioner should add her share to the effort to sustain national security despite her critical views and her ideological perception of what a just national policy is.'[25] No recommendation was added regarding accommodating her conscience by assigning her to posts that were not in direct contact with acts that sustained the occupation. Why did Justice Procaccia take the trouble to recommend accommodating Shvili's conscience if she thought that he was a selective conscientious objector and if she thought that selective conscientious objection should not be tolerated? Shvili's request was not accepted in full, and therefore he was not tolerated by the court, but at the same time, the court offered his position a certain degree of sympathy. It seems that despite Justice Procaccia's strict rhetoric objection to selective conscientious objection, and despite her unwillingness to differentiate between 'institutional selective conscientious objection' and 'ideological selective conscientious objection', she did in fact differentiate between the two after all. The informal, 'recommended' remedy that she gave to Shvili (and did not give to Zonshein and Milo) leads to a conclusion that Justice Procaccia does think that the Shvili case gives rise to a special and more tolerable kind of selective conscientious objection.

Second, two of the three judges in the Shvili case (Justice Chayut and Justice Yoram Danziger) took the trouble to add an obiter according to which, and in future cases, contingent pacifism, which is in fact 'selective' conscientious objection, may justify granting conscientious exemptions from army enlistment. As far as obiter dicta go, this could be the beginning of a new thinking and a new practice in the Israeli Supreme Court regarding cases of selective conscientious objections. It is not at all clear, however, why this approach was not applied in the Shvili case itself.

Conclusion

Most democratic states tolerate, to various extents, conscientious objectors. The same states do not tolerate, almost completely, what they perceive as selective conscientious objection. This is disturbing for two main reasons. First, it is not at all clear whether there is meaningful conceptual difference between 'selective'

25 *Shvili*, para 16 to Justice Procaccia's decision.

conscientious objection and 'absolute' conscientious objection. Second, even if such a difference exists it bears no significant moral or practical implications. There are no moral or prudential prevailing reasons that justify treating both types of objection differently. This chapter maintained that the argument about numbers, the argument about solidarity or unity and the evidential problem cannot justify treating these types of objection differently. It was also argued that the 'lack of choice' argument which justifies tolerating conscientious objections equally applies to selective conscientious objection and to non-selective conscientious objection. Finally, it was argued that recent Israel Supreme Court's decisions have failed to deal justly and fairly with the issue of selective conscientious objection.

Generally speaking, the time has come for state organs to be persuaded by current philosophical thinking and to abandon old and misguided conceptions and calcifications regarding the issue of (selective) conscientious objection. Within the Israeli context there is still much work to be done by Israeli Supreme Court. The Court must rethink and re-evaluate its approach towards the distinction between selective and non-selective conscientious objections; the distinction between female and male objectors to enlisting into the army; and the distinction between secular-humanist and religious conscientious objectors. In this chapter I offered initial and partial suggestions with regard to the conceptual, moral and practical distinction between selective and non-selective conscientious objections. It would be safe to assume that Israeli Supreme Court will have the opportunity to engage with this issue yet again and presumably quite shortly. Hopefully, the court will take this opportunity to deviate from its recent erroneous decisions and to deliver a decision that may lead the way to a longed for legal change in this issue.

Chapter 10
Claims for Refugee Protection in Canada by Selective Objectors: An Evolving Jurisprudence

Yves Le Bouthillier

In the last few years, a number of US soldiers have refused to serve in armed conflicts in Afghanistan and Iraq and have claimed refugee protection in Canada to escape punishment by US military law. They argue that these wars are contrary to international law and/or are conducted in violation of international humanitarian law. In other words, their objection is 'selective' as opposed to a general refusal to participate in all armed conflicts because of deeply held religious, moral or ethical beliefs.

Given that these claims are made by soldiers from a democratic state, and, more particularly, Canada's closest neighbour, it is not surprising that they have attracted much public attention.[1] In the past, the vast majority of these claims were made by selective objectors from countries with no democratic tradition or with dubious democratic credentials.[2]

An early seminal case on this issue is the Federal Court of Appeal decision in 1993 in *Zolfagharkhani*[3] regarding the claim by a selective objector from Iran. In that case, Justice MacGuigan, in distinguishing the former leading authority on the topic, *Musial*,[4] opened the door somewhat to refugee claims based on selective objections.[5]

---

1 In one of the decisions concerning a selective objector from the United States, *Hinzman v Canada (Minister of Citizenship and Immigration)*, 2009 FC 415 [*Hinzman 2009*], the Federal Court referred to the public interest in the case at para.97: 'If Court room attendance at the hearing is anything to go by, this application has attracted significant public interest and debate.'

2 The courts have reviewed cases from many countries, including Algeria, Armenia, El Salvador, Federal Republic of Yugoslavia, Iran, Iraq, Israel, Lebanon, Mexico, Russia, Serbia, Turkey, Venezuela and Yemen.

3 *Zolfagharkhani v Canada (Minister of Employment and Immigration)*, [1993] 3 FC 540 (FCA) [*Zolfagharkhani*].

4 *Musial v Canada (Minister of Employment and Immigration)*, [1982] 1 FC 290 (FCA) [*Musial*].

5 Edward C. Corrigan, 'Refusal to Perform Military Service as a Basis for Refugee Claims in Canada', *Immigration Law Reporter* (2000), notes also a few decisions of the Immigration and Refugee Board in the 1980s which questioned the restrictive interpretation

In his analysis, he referred[6] to paragraph 171 of the UNHCR *Handbook on Procedures and Criteria for Determining Refugee Status* which states that:

> 171. Not every conviction, genuine though it may be, will constitute a sufficient reason for claiming refugee status after desertion or draft-evasion. It is not enough for a person to be in disagreement with his government regarding the political justification for a particular military action. Where, however, the type of military action, with which an individual does not wish to be associated, is condemned by the international community as contrary to the basic rules of human conduct, punishment for desertion or draft evasion could, in the light of all other requirements of the definition, in itself be regarded as persecution.[7]

Since then, both the Federal Court and the Federal Court of Appeal have been called upon to look in some depth at the issue of objectors in the context of judicial reviews of decisions of the Refugee Protection Division (RPD) of the Immigration and Refugee Board of Canada.[8] In 2008, a study reached the conclusion that the jurisprudence on this topic 'has been plagued by fragmentation, incoherence and inattention.'[9] However, as this chapter will demonstrate, a number of recent decisions have attempted to clarify the scope and content of paragraph 171 as well as a number of other questions that are raised by these types of claims.

Distinguishing Between an 'absolute conscientious objector' and a 'selective objector'

In the *Zolfagharkhani* decision, Justice MacGuigan does not use the term 'selective objectors' nor does he expressly distinguish these type of claims from 'absolute conscientious objectors'. In a recent decision, *Lebedev*, Justice de Montigny notes that the use of 'conscientious objector' as an all-encompassing

in *Musial*. See http://www.edcorrigan.ca/immigration-services.html accessed 30 January 2013.

6 *Zolfagharkhani*, p. 555.

7 UNHCR, *Handbook on Procedures and Criteria for Determining Refugee Status under the 1951 Convention and the 1967 Protocol relating to the Status of Refugees*, (Geneva: UNHCR, 1992) [*Handbook*].

8 This chapter will only consider decisions of the Federal Court of Canada and the Federal Court of Appeal of Canada. However, there have been many more decisions on this issue by the Refugee Protection Division (RPD) of the Immigration and Refugee Board (IRB) and its predecessor, the Convention Refugee Determination Division as noted by Martin Jones, 'The Refusal to Bear Arms as Grounds for Refugee Protection in the Canadian Jurisprudence', *International Journal of Refugee Law*, 20/1 (2008), p. 125 [*Jones*].

9 Jones, p. 126. Federal Court Judge Yves de Montigny expressed similar views in 2007 in the decision *Lebedev v Canada (Minister of Citizenship and Immigration)*, 2007 FC 728 [*Lebedev*], para.2.

term in *Zolfagharkhani* created some ambiguity as to the grounds on which it was decided.[10] He also observes that the factual situation described and analysed by MacGuigan reveals different types of objectors to armed conflicts. Zolfagharkhani was an Iranian soldier who took an active part in combat during the war against Iraq. He was not opposed to fighting in all armed conflicts and as such could not be considered an 'absolute conscientious objector'. After his discharge from the army, he was called upon once more to serve as an army paramedic in a war against the Kurdish movement. In his last week of training for this position he discovered that his government intended to resort to chemical weapons. Opposed to the use of these weapons, he fled Iran. On the basis of the facts, he fits the description of a 'selective objector' and, implicitly, that is how MacGuigan understood his claim for refugee status as he noted that 'no issue is raised as to conscientious objection in relation to war in general ... The issue as to conscientious objection relates solely to participation to chemical warfare.'[11]

In *Lebedev*, Justice de Montigny dissipates any ambiguities by distinguishing clearly between two types of objectors: 'As I see it, conscientious objection applies to those who are totally opposed to war because of their politics, ethics or religion. Selective objection really refers to cases in which an applicant opposes a war he feels violates international standards of law and human rights.'[12]

While the extent to which one is opposed to an armed conflict differs, in both cases a claimant for refugee protection has to convince decision makers that his or her opposition is genuine. However, as de Montigny explains, the decision maker will also have to assess carefully the objective nature of the selective objector's claim.

> The first type of claim, conscientious objection, raises subjective issues. Decision-makers must evaluate the applicant's personal beliefs and conduct to see if his claim is genuine. The second type of claim requires both a subjective and objective assessment of the facts. Along with evaluating the sincerity of an applicant's beliefs, a decision-maker must look at whether the conflict objectively violates international standards. The two types of objections should be treated as distinct categories – just as they are distinguished in paragraphs 171 and 172 of the UNHCR Handbook.[13]

---

10 *Lebedev,* para.42.

11 *Zolfagharkhani*, p. 553.

12 *Lebedev*, para.44.

13 *Lebedev*, para.45. Paragraph 172 of the *Handbook* regarding conscientious objectors reads as follows: 'Refusal to perform military service may also be based on religious convictions. If an applicant is able to show that his religious convictions are genuine, and that such convictions are not taken into account by the authorities of his country in requiring him to perform military service, he may be able to establish a claim to refugee status. Such a claim would, of course, be supported by any additional indications that the applicant or his family may have encountered difficulties due to their religious

A recent Federal Court of Appeal decision, *Ates*, has taken a very restrictive view of claims based on an absolute conscientious objection. The claimant, a Kurd from Turkey, refused to report for compulsory military service. In that country it is not possible for pacifists to opt for alternative service such as a non-combat position or an assignment in the civil service. If returned to Turkey, he would likely be jailed.[14] The Federal Court upheld the RPD decision to reject this claim:

> The Board member accepted that Mr. Ates was a pacifist and that his refusal to serve in the armed forces violated Turkish law, as all young men are subject to conscription. However, it was also found that he would not suffer disproportionately severe punishment on account of his race or religion.[15]

Ates could not succeed on the sole basis that he was an absolute conscientious objector. He would have had to meet the criteria found in paragraph 169 of the *Handbook*:

> 169. A deserter or draft-evader may also be considered a refugee if it can be shown that he would suffer disproportionately severe punishment for the military offence on account of his race, religion, nationality, membership of a particular social group or political opinion. The same would apply if it could be shown that he has a well-founded fear of persecution on these grounds above and beyond the punishment for desertion.[16]

As Ates could not demonstrate that his imprisonment was disproportionately severe punishment, his claim did not succeed.

The result was similar at the Federal Court of Appeal as the court answered in the negative the following question without providing any reasons:

> In a country where military service is compulsory, and there is no alternative thereto, do repeated prosecutions and incarcerations of a conscientious objector for the offence of refusing to do his military service, constitute persecution based on a Convention refugee ground?[17]

---

convictions.' While this paragraph suggests that a claim of a conscientious objector must be based on 'religion', de Montigny takes a broader view, stating at para.46 of *Lebedev* that 'this notion should be expanded to recognize that moral principles may also be, for a number of people, sufficiently compelling to ground and organize their lives.'

14 *Ates v Canada (Minister of Citizenship and Immigration)*, 2004 FC 1316 [*Ates*], para.17.

15 *Ates*, para.16.

16 *Handbook*, note 7.

17 *Ates v Canada (Minister of Citizenship and Immigration)*, 2005 FCA 322 [*Ates FCA*].

The Court closed the door to absolute conscientious objectors even in cases of 'repeated prosecutions and incarcerations'. The case was subsequently and frequently referred to by the Federal Court to reject claims based on opposition to all wars or some wars which do not fall within the ambit of paragraphs 171 or situations provided for in paragraph 169.[18]

However, in one case the Court also suggested that the law could evolve in the future so as to recognize claims by conscientious objectors:

> Given that conscientious objection is a fundamental aspect of the right to freedom of thought, conscience and religion enshrined in international instruments such as the *Universal Declaration of Human Rights* and the *European Convention on Human Rights*, it may be that as the law continues to evolve in this area, both on the international and domestic fronts, a sincerely-held political or religious objection to a specific war may some day provide a sufficient basis on which to ground a claim for refugee protection. This, however, represents the 'international consensus of tomorrow' ... and not the state of the law today.[19]

Ordinary Laws of General Application: From a Restrictive to a More Liberal Approach

The 1982 Federal Court of Appeal decision in *Musial*[20] was subsequently interpreted as barring any refuge claim by objectors on the basis that a government sanctioning soldiers for deserting the army is prosecuting them for transgressing 'an ordinary law of general application' and not engaging in persecution.[21] The argument is that the claimant does not fear a serious violation of his or her human rights. He or she simply wants to avoid a penalty provided by law.

In *Zolfagharkhani*, Justice MacGuigan rejected this absolute bar to a refugee claim, noting that 'any given ordinary law of general application in a dictatorial or totalitarian state may well be an act of political oppression.'[22] While a claimant must have a subjective fear, the key factor is that a state perceives a soldier's

---

18 *Hinzman v Canada (Minister of Citizenship and Immigration)*, 2006 FC 420 [*Hinzman 2006*], para.225. See also *Ielovski v Canada (Minister of Citizenship and Immigration)*, 2008 FC 739 [*Ielovski*]; *Loshkariev v. Canada (Minister of Citizenship and Immigration)*, 2006 FC 670; *Treskiba, v Canada (Minister of Citizenship and Immigration)*, 2009 FC 15.

19 *Hinzman 2006*, para.233.

20 In *Musial*, the claimant refused to serve in the Polish army for fear of being assigned to Afghanistan in support of the Soviet Union invasion of that country. The basis for this claim was that both Poland and the USSR were part of the Warsaw Pact. Chief Justice Thurlow, concurring in the case, mentioned at p. 291 that that there was only 'a bare possibility' that the claimant would be required to serve in Afghanistan.

21 *Zolfagharkhani*, p. 549.

22 *Zolfagharkhani*.

refusal to serve as an expression of his or her political opinion and prosecutes him or her on that basis. In that case a state could be considered a persecutor.[23]

Even if MacGuigan made it possible for a claimant to demonstrate that a state is using an 'ordinary law of general application' as a persecutory tool, the claimant's evidentiary onus remains heavy. His comments, arguably based on the facts before him, link the relativity of an ordinary law of general application to a 'dictatorial or totalitarian state'. These were seemingly the states he had in mind when he wrote that this 'absolute proposition of prosecution ... could not be supported in relation to the majority of countries from which refugee cases arise.'[24] He adds later in the decision that laws are presumed at the outset to be valid and neutral, 'even in non-democratic societies'.[25] Therefore, in all cases, it is for the claimant to 'show that the laws are either inherently or for some other reasons persecutory.'[26]

States, even dictatorial ones, will rarely adopt laws that are, on their face, persecutory. In most cases, it is the application of the law to a specific claimant that results in persecution. For instance, in *Zolfagharkhani*, the law was applied to all army evaders independently of their reasons for refusing to serve. The expected application of this law to the claimant despite his political opposition to certain aspects of the armed conflict rendered its application persecutory with regard to him. As stated by MacGuigan: 'the ordinary Iranian conscription law of general application, *as applied in a conflict in which Iran intended to use chemical weapons*, amounts to persecution for political opinion'.[27]

If, as mentioned above, there is a presumption that all laws are valid and neutral, the decision in *Zolfagharkhani* nevertheless represents a significant shift as it did away with a barrier that previously appeared to many as insurmountable.[28]

Degree of Involvement Required

A selective objector can be opposed to a conflict because he or she considers that the conflict is contrary to international law norms even if the objector would not be directly involved in the violation of these norms. For instance, a soldier can be

---

23 The Supreme Court of Canada in *Canada (Attorney General) v Ward*, [1993] 2 SCR 689 [*Ward*], at p. 747, confirmed that when looking at a claim of persecution, the state of mind of the persecutor is a key factor.

24 *Zolfagharkhani*, p. 549.

25 *Zolfagharkhani*, p. 552.

26 *Zolfagharkhani.* Another avenue for a claimant is to demonstrate that the law under which he would have been prosecuted is illegal. See *Mohamed v Canada (Minister of Employment and Immigration)*, (1994) 176 National Reporter 60 (FC). It would only be in rare instances that one would be able to demonstrate that a draft evading law was adopted illegally.

27 *Zolfagharkhani*, p. 555 (my emphasis).

28 See *Ciric v Canada (Minister of Employment and Immigration)*, [1994] 2 FC 65 (TD) [*Ciric*], p. 72 where the applicant notes this shift.

of the view that the use of force by his or her government amounts to an act of aggression, and/or will involve violation of international humanitarian law. For the courts these opinions do not justify a claim of persecution. Selective objectors also need to demonstrate, based on paragraph 171 of the *Handbook*, that they could have been involved in acts that are contrary to basic rules of conduct. However, Courts have also recognized that if the claimant holds a senior position within the military, a claim based on a war's illegality could be successful.

In *Zolfagharkhani* Justice MacGuigan examines the 'applicant's prospective degree of participation.'[29] In doing so, he suggests that the claimant has to demonstrate a potential association with a contravention of 'basic rules of human conduct' and not just an opposition to the use of chemical weapons. He emphatically rejects the Board's finding that 'the appellant, as a paramedic, would not be a combatant at all, but perform merely a Florence Nightingale role of "helping his brothers on both sides of the camp."'[30] In his view, Zolfagharkhani's role as a paramedic would implicate him directly in chemical warfare. He would be instrumental in facilitating their use by treating Iranian soldiers inadvertently hurt by these weapons and the injured Kurdish prisoners that would need to be questioned.[31] It is Zolfagharkhani's role during the conflict that creates for Justice MacGuigan 'an issue of conscience'.

However, other parts of the decision suggest that, given the nature of the conflict, any association with it could amount to a contravention of 'basic rules of human conduct':

> I have already indicated that I find the Board's reasoning untenable, because the distinction it drew between participation and non-participation in military activity is unrealistic and even naive, in the context of the horrendous nature of chemical weapons, which by their nature make no distinction between combatant and non-combatant, or sometimes even between attacker and defender, all being potential victims.[32]

Based on the above, the decision left some uncertainty as to the 'prospective degree of participation required', if any, to justify a claim for refugee status in this type of situation.

This issue has since been expressly addressed in *Hinzman*.[33] The claimant, a US soldier, argued that it would be sufficient for him to demonstrate that the war in Iraq was in violation of international law to fall within the ambit of paragraph 171 of the *Handbook*. It would place too heavy a burden on him to also have

29 *Zolfagharkhani*, p. 547.
30 *Zolfagharkhani*.
31 *Zolfagharkhani*, p. 547.
32 *Zolfagharkhani*, p. 553.
33 *Hinzman 2006*.

to demonstrate that he would have been involved in unlawful acts.[34] Justice Mactavish rejected this argument. Distinguishing between a 'foot soldier' and a 'senior officer', she explained that for the former, his or her claim had to be based on conduct 'on the ground'.[35]

According to Justice Mactavish, selective objectors under paragraph 171 are those who, had they not deserted, could have been involved in actions on the ground that would have led to their exclusion from the *Refugee Convention.*[36] Article 1(F)(a) of the *Convention* excludes from refugee protection an individual for whom there are serious reasons to believe he or she 'has committed a crime against peace, a war crime, or a crime against humanity, as defined in the international instruments drawn up to make provision in respect of such crimes.'[37] If, during or after an armed conflict, a soldier claims refugee protection, such claim will be rejected if he or she was involved in one of the illegal acts listed in Article 1 F (a). In *Zolfagharkhani*, Justice MacGuigan had raised the possibility that if the paramedic had participated in chemical warfare, he could have been excluded from refugee protection.[38] In the case of *Hinzman,* Justice Mactavish dismissed the possibility that the claimant, a mere foot soldier, would have been excluded for a 'crime against peace' on the basis that his state was involved in an illegal war.[39] As the illegality of the war (*jus ad bellum*) is not relevant to his claim, there is no need for the RPD or the Court to assess the legality of the conflict itself.[40]

34 *Hinzman 2006*, para.4.

35 *Hinzman 2006*, para.166. See also *Radosevic v Canada (Minister of Citizenship and Immigration)*, [1995] FCJ No 74 (QL) (TD).

36 *Hinzman 2006*, para.150.

37 189 UNTS 150, Can TS 1969 No 6 (entered into force April 22, 1954). The regime of exclusion in the Convention was incorporated into Canadian law. See the *Immigration and Refugee Protection Act*, SC 2001, c 27 [IRPA], s 98. Sections E and F of Article 1 of the *Convention* are reproduced verbatim at the end of the Act.

38 *Zolfagharkhani*, p. 548. This has been described as the 'right not to be a persecutor' in *Hinzman 2006*, para.119, referring to Mark R. von Sternberg, *The Grounds of Protection in the Context of International Human Rights and Humanitarian Law: Canadian and United States Case Law Compared* (The Hague; New York: Martinus Nijhoff, 2002), pp. 124 and133. Another early case on this issue is *Diab v Canada (Minister of Employment and Immigration)*, [1994] FCJ No 1277 (QL) (FCA).

39 *Hinzman 2006*, para.188.

40 *Hinzman 2006*, para.165. She distinguished the decision of the Federal Court of Appeal in *Al-Maisri v Canada (Minister of Employment and Immigration)* (1995), 183 National Reporter 234 (FCA) [*Al-Maisri*]. In that case the Court had accepted the claim from a draft evader from Yemen opposed to the legality of the invasion of Kuwait by Iraq, an invasion supported by his country. Justice Mactavish found that *Al-Maisri* had not dealt directly with the issue raised in *Hinzman.* See para.148. However, she did certify a question to the Federal Court of Appeal on the basis that the matter was not entirely free of doubt because of the decision in *Al-Maisri. See Hinzman 2006*, para.238.

In *Lebedev*,[41] Justice de Montigny expressed some reservations about limiting claims by selective objectors to those who would otherwise have been excluded under section 1F(a). He endorses Martin Jones's warning against an automatic equivalence between the law of exclusion of persecutors and the law of inclusion of soldiers who object in order not to become persecutors. With regard to the former, Jones explains that the concept of 'participation' has been interpreted restrictively given the seriousness of a finding of exclusion.[42] As mentioned by Justice de Montigny, the test for exclusion requires that 'an applicant knew of an organization's crimes and shared its purpose in committing them.'[43] In his view, this is a more demanding test than the need to demonstrate 'association' with such acts as provided for in paragraph 171 of the *Handbook*.[44] He sees no reason to apply the more restrictive test of 'participation' to claims of inclusion by selective objectors.[45] Instead, the RPD and the Court should be more open and flexible when assessing the potential involvement of selective objectors 'on the ground'. He suggests that this approach explains Judge MacGuigan's finding in *Zolfagharkhani*:

> After all, the Federal Court of Appeal was able to find that a paramedic's role in treating injured soldiers was sufficient to bring him within the purview of paragraph 171 of the UNHCR Handbook in *Zolfagharkhani*, above. That case clearly stands as an indication of how we should approach the difficult moral dilemma confronted by those called to serve in wars of dubious legitimacy.[46]

The approach put forward by Justice de Montigny was adopted a year later by Justice Barnes in *Key*.[47] In this case, the claimant is a US soldier assigned to security duties in Iraq. He was involved in over seventy night-time raids of private Iraqi homes in search of weapons.[48] The RPD concluded that while these acts involved a 'disturbing level of brutality' and could be contrary to a number of provisions found in international conventions on humanitarian law, they did not reach the level needed to be excluded as war crimes or crimes against humanity.[49] It ruled that they could not justify a claim for refugee protection based on paragraph 171 of the *Handbook* since 'refugee status can only be conferred where a soldier's

41 *Lebedev*.

42 Jones, pp. 143-4.

43 Lebedev, para.87. See *Ramirez v Canada (Minister of Employment and Immigration)*, [1992] 2 FC 306 (FCA).

44 *Lebedev*, para.86.

45 *Lebedev*, para.88.

46 *Lebedev*, para.89. However, one should note that Justice MacGuigan also mentions that the paramedic could have been excluded. In other words, he appears to have a broad view of 'participation' whether in the context of 'inclusion' or 'exclusion'.

47 *Key v Canada (Minister of Employment and Immigration)*, 2008 FC 838 [*Key*].

48 *Key*, para.6.

49 *Key*.

past combat experiences or the expectations for further combat service would constitute excludable conduct under the *Convention Relating To The Status Of Refugees*.'[50]

Justice Barnes noted the difference of terminology between exclusions (requiring 'participation') and paragraph 171 (requiring 'association').[51] He found no compelling reasons to limit violations of 'basic rules of conduct' in paragraph 171 to crimes against humanity and war crimes.[52] He concluded that

> [...] officially condoned military misconduct falling *well short* of a war crime may support a claim to refugee protection. Indeed, the authorities indicate that military action which systematically degrades, abuses or humiliates either combatants or non-combatants is capable of supporting a refugee claim where that is the proven reason for refusing to serve.[53]

Whether one looks at the issue on the basis of 'association' or 'participation', the approach that an army takes to violations of fundamental human rights will be an important factor in determining whether a selective objector could have been involved in actions contrary to basic human conduct. One can imagine a number of scenarios:

- If an army is considered as an organization 'principally directed to a limited, brutal purpose'[54] mere membership in it could trigger the exclusion regime.[55] A claim by a selective objector would likely be accepted if he could demonstrate that he left at the first occasion he had.
- If an army, without falling within the definition of a 'brutal organization', nevertheless commits systematic violations of fundamental human rights in a specific conflict and that the higher authorities within the army, by their conduct, condone these actions, the claim for refugee protection would likely succeed. In Hinzman, the claimant argued that that if there is a high number of violations on the ground and a climate of tolerance by the upper ranks, soldiers are more likely to be implicated in situations involving violations of international humanitarian law. Both the RPD and the Court did not accept that there was either systemic violation of international

50 *Key*, para.14.

51 *Key*, para.21.

52 *Key*, para.27.

53 *Key*, para.29 (my emphasis). However, see *Gebre-Hiwet v. Canada (Minister of Employment and Immigration)*, 2010 FC 482 [*Gebre-Hiwet*] where the Court links the refusal to take part in a conflict to being forced to commit a crime against humanity.

54 See *Ramirez*.

55 In *Hinzman 2006*, para.154, the Court did mention that there was no suggestion that the US army met that description.

humanitarian law or that the US Army condoned such conduct in Iraq.[56] If the Court had reached a contrary conclusion on the facts, one assumes that Hinzman's claim could have been accepted as the Court added: 'the availability of refugee protection should be limited to deserters from armed conflicts where the level and nature of the conflict, and the attitude of the relevant government, have reached a point where combatants are, or may be, required, on a sufficiently widespread basis, to breach the basic rules of human conduct.'[57]

- If an army is involved only sporadically in violations of international humanitarian law and violations are investigated and punished, the claim for refugee protection would likely be rejected, unless the claimant establishes that he or she was at risk of being involved in these violations. In *Hinzman*, the RPD accepted that some members of the US army had violated international humanitarian law.[58] The Court, observing that there is never any guarantee that a war will be free of atrocities,[59] ruled that isolated incidents are not sufficient to justify a selective objector's claim.[60]

In *Hinzman,* the Court agreed with the RPD that the claimant failed to establish that he would have been called to violate international humanitarian law while deployed in Iraq.[61]

When the claim is by a foot soldier, it is relevant to examine what task he or she was normally expected to perform during service. While in *Zolfagharkhani* a paramedic was found to be a 'participant' in the context of a chemical warfare, the conclusion could have been different for other types of conflict.[62] In *Ielovski*, the claimant had previously been posted as a mechanic in a military air base and feared that he would be called to serve again. Justice de Montigny found that 'Based on the duties he performed during his military service, we may assume

---

56 *Hinzman 2006*, para.187. For other cases where the claimant did not establish systemic violations of human rights see *Sounitsky v Canada (Minister of Citizenship and Immigration)*, 2008 FC 345; *Mohilov v Canada (Minister of Employment and Immigration)*, 2008 FC 1292.

57 Hinzman 2006, para.170.

58 *Hinzman 2006*, para.176. The Board had also found 'that incidents of human rights violations by American military personnel were investigated, and the guilty parties punished.'

59 *Hinzman 2006*, para.169.

60 See also *Popov v Canada (Minister of Employment and Immigration)*, (1994); *Volkovitsky v Canada (Minister of Citizenship and Immigration)*, 2009 CF 893.

61 *Hinzman 2006*, para.189.

62 In *Colby v Canada (Minister of Citizenship and Immigration)*, 2008 FC 805, a US soldier serving as a medic in Iraq argued that he was ordered not to use anaesthetic when treating Iraqi combatants. The Court did not assess whether these acts fell within paragraph 171 of the *Handbook* as the claim was dismissed because he had not made sufficient efforts to request state protection.

that he would not have been posted to combat operations. It is therefore far from evident that the applicant would personally have been called on to directly or indirectly participate in international law violations.'[63]

If a foot soldier cannot claim refugee status for having opposed a violation of the *jus ad bellum* (crimes against peace), courts have acknowledged that a senior military officer can if that official would otherwise have been asked to play a role in the planning or triggering of the war:

> A refusal to be involved in the commission of a crime against peace could indeed potentially bring a senior member of a government or military within the ambit of paragraph 171 ... This presupposes, however, that the involvement and level of the individual is such that he or she could be guilty of complicity in a crime against peace ... it is only those with the power to plan, prepare, initiate and wage a war of aggression who are culpable for crimes against peace.[64]

Finally, even if a claimant establishes that his or her degree of involvement in actions 'contrary to the basic rules of human conduct' could have been sufficient to justify a claim, the RPD will still have to consider the nature of the punishment that he or she would have incurred for deserting or evading draft. This issue is addressed later in this chapter.

Whether there is a Need for Condemnation by the International Community?

Paragraph 171 of the UNHCR *Handbook* provides that the type of military action with which an individual does not want to be associated is one that 'is condemned by the international community'.[65] Some claims have been rejected on the basis that no evidence of international condemnation had been proven:

> It is not enough to submit evidence with a view to establishing that the actions of the Israeli army with respect to the Palestinians are bad or even deplorable ... it must be established that the action is condemned by the international community. The applicants did not establish this fact.[66]

This requirement, if it is indeed one, raises a number of issues. What is the international community? Who speaks for it? What evidence is sufficient to establish condemnation by the international community? A statement by the UN Security Council or the General Assembly, firmly condemning a state's

---

63 *Ielovski*, para.10. See also *Treskiba*.

64 *Hinzman 2006*, paras 141-2. See also *Lebedev*, para.85.

65 *Handbook*.

66 *Zuevich v. Canada (Minister of Citizenship and Immigration)*, (1999) 172 FTR 1 (TD), para.41.

actions, could be sufficient.[67] However, given the potential for stalemate within this political and complex organization, Courts have rejected the need for formal condemnation by the UN:

> It must be remembered that this world organization, intent on maintaining peace, must act of necessity slowly and carefully if it is to remain the honest broker in any conflict. Fortunately, respected organizations like Amnesty International, Helsinki watch and ICRC, are able to move quickly, study sufficiently and make pronouncements. And all did so here which surely the board should have seen as condemnation by the world community.[68]

In *Lebedev,* concerning a Russian soldier's refusal to fight in Chechnya, an international condemnation by the Security Council was not imaginable given that Russia can use its veto to block the adoption of a resolution. In any event, Justice de Montigny stated that a condemnation by the UN was not a prerequisite to the application of paragraph 171:

> Conversely, there will also be instances where political expediency will prevent the UN or its member states from condemning massive violations of international humanitarian law. This is why reports from credible non-governmental organizations, especially when they are converging and hinge on ground staff, should be accorded credit. Such reports may be sufficient evidence of unacceptable and illegal practices.[69]

However, Justice de Montigny went further, stating that the real focus of the inquiry has to be whether there have been breaches of international law or international humanitarian law. International condemnation would be evidence of such violations but would not be a prerequisite to the application of paragraph 171.[70]

Referring to Russia's actions in Chechnya, Justice de Montigny also rejected the notion that explicit evidence of that state's intent was required to establish that it was engaged in systematic violations of human rights.

---

67 Although, with regard to the General Assembly, a majority vote with some regions of the world opposing a resolution or abstaining would carry less weight.

68 *Ciric*, p. 217.

69 *Lededev*, para.70. In a subsequent case, *Ielovski*, para.10, Justice de Montigny found that evidence of international condemnation was lacking as the claimant had relied only on a few newspaper and magazine articles.

70 See also *Hinzman 2006*, para.134. See also *Tewelde v. Canada (Minister of Citizenship and Immigration)*, (2007) 65 Immigration Law Reporter (3d) 267 (FC). See also *Key*, para.21: 'the response of the international community to the legitimacy of a particular conflict or to the means by which it is being prosecuted has generally been seen as relevant but not a determinative consideration'.

> It would set a dangerous precedent to accept that Russia had not systemically violated human rights solely because it had not admitted to it directly. Massive human rights violations may take place not only through deliberate policy, but also through official indifference or by being condoned by the authorities. Transgressions of international norms should always be taken into account in assessing a refugee claim, however they come about.[71]

With these clarifications, Judge de Montigny reaffirmed that serious human rights violations are at the core of refugee determination, that evidence of these violations can take various forms, and that they are not limited by restrictive interpretations of 'international condemnation' and 'intent'.

Is the Nature of the Punishment Persecutory?

Punishment for draft evaders and deserters varies from country to country, both in terms of what the legislation provides for and in practice. In some cases, there could be very little risk of punishment or punishment could be minor,[72] while in others the risk of long term imprisonment or even execution is the more likely outcome.[73] In many other cases, it can vary from a fine to long term imprisonment. In addition, the practice within a state can change over time. There is evidence that some countries, for instance, will prosecute more deserters and increase the severity of punishment as a war continues and there is a need to recruit and retain more soldiers.[74]

71 *Lebedev*, para.78.

72 See for instance *Budaghyan v Canada (Minister of Citizenship and Immigration)*, 2002 FCT 20, para.8. See also *Hinzman v Canada (Minister of Citizenship and Immigration)*, 2007 FCA 171 [*Hinzman FCA 2007*], where the Federal Court of Appeal notes at pp. 25-6, para.48: 'while punishment for desertion can include imprisonment, the evidence indicates that the vast majority of Army deserters in the United States have not been prosecuted or court-martialled. Rather, approximately 94% of deserters have been dealt with administratively and merely receive a less-than-honourable discharge from the military.' These findings were however challenged in subsequent cases.

73 In *Ciric*, the claimants argued that if they were returned to their country (Serbia), they could either be imprisoned or executed. The Federal Court found in favour of the claimant on this issue (p. 78): 'It is just impossible to conceive that the board could accept in the most vicious of civil wars that the only punishment this couple would receive is a fine. The source itself is suspect – a federal government official.'

74 See Paul von Zielbauer, 'Army is Cracking Down on Deserters', *The New York Times*, 9 April 2007 [Zielbauer]. This article was referred to in *Glass v Canada (Minister of Citizenship and Immigration),* 2008 FC 881 [*Glass*], where the Court mentioned at para.12 that: 'The evidence emanating from US Army sources shows that prosecution for desertions were increasing, with prison terms imposed, particularly against those members who had publicly denounced the war in Iraq.'

A number of decisions have considered the severity of the anticipated punishment to determine if it amounts to persecution. In doing so, they have confirmed, as provided in paragraph 167 of the *Handbook*,[75] that punishment in itself is not persecutory and they have considered, instead, whether, in the circumstances of each case, a claimant would, as provided in paragraph 169, 'suffer disproportionately severe punishment for the military offence on account of his race, religion, nationality, membership of a particular social group or political opinion.'[76] In a number of cases the Federal Court agreed with the RPD that length of imprisonment, even when exceeding a year, did not amount to disproportionate punishment.[77] In looking at the punishment, the Court also considers whether the conditions of imprisonment meet international standards.[78]

In *Hinzman,* the claimant argued that he would likely face between one to five years in a military prison, have his pay forfeited, be dishonourably discharged and face some employment and societal discrimination. He also argued that he would likely be treated more harshly than other deserters as the American authorities would want to dissuade similar refugee claims in the future.[79] While the Court did not rule as to whether this would amount to disproportionate punishment,[80] the RPD's view was that this would not be 'grossly disproportionate to the inherent seriousness of the offence of desertion'.[81] It also did not accept that he would be treated more harshly because he had made a refugee claim.[82]

---

75 *Handbook.*

76 *Handbook* para.169. The RPD, in addition to considering whether the punishment could amount to persecution under one of the grounds enumerated in the *Refugee Convention,* as provided for in Section 96 of the IRPA, will also determine if the punishment would justify protection under Section 97. This section provides protection to persons that can establish that, if returned to their country, they could be tortured, their life would be at risk or they could suffer cruel and unusual treatment or punishment. Under that section, there is no need to establish a *nexus* between the grounds enumerated in the Convention and the treatment or punishment.

77 See *Volovich v Canada (Minister of Citizenship and Immigration),* 2004 FC 269 (two months' imprisonment for an evading soldier was reasonable); *Ozunal v Canada (Minister of Citizenship and Immigration),* 2006 FC 560 (imprisonment of up to three years is not 'excessive or over-harsh sentence'); *Sounitsky* (a possibility of imprisonment for up to 56 days is not so extreme as to be cruel and unusual); *Lowell v Canada (Minister of Citizenship and Immigration),* 2009 FC 649 (possibility of 15 months imprisonment); *Kirkoyan v Canada (Minister of Citizenship and Immigration),* 2011 FC 1217 (three to six months in prison is not disproportionate or persecutory). See however *Glass* para.34 where the Justice seems to suggest that an imprisonment for five years would be excessive. Note that the Justice also considered that the conduct of the US army in Iraq may be considered a violation of basic human conduct under para.171.

78 See *Ozunal* at para.29.

79 *Hinzman 2006,* paras 79 and 85-6.

80 *Hinzman 2006,* para.226.

81 *Hinzman 2006,* para.83.

82 *Hinzman 2006,* para.82 See however *Rivera v Canada (Minister of Citizenship and Immigration),* 2009 FC 814 where the Federal Court returned a Pre-Removal Risk Assessment for another determination as the immigration officer had failed to assess, as argued by the

If a claimant can establish that his or her objection is founded on actions that have been condemned by the international community as contrary to basic human rules of conduct (paragraph 171 of the Handbook), would any punishment amount to persecution?[83] Martin Jones suggests that the Court will still consider the punishment, but will balance it with the seriousness of the situation that the claimant was confronted with:

> There is also some suggestion that *any* punishment for refusal to participate in this type of prohibited military service will constitute persecution. However, a more conventional approach would be to consider whether the punishment in question is a serious violation of the claimant's human rights. Given the context of punishment for failing to comply with prohibited conscription – the punishment of an essentially 'innocent' man for failing to comply with an illegal act – it is likely that anything more than brief imprisonment would constitute persecution.[84]

In some countries, the real risk to the claimant is not what is found in ordinary legislation but extra-judicial punishment, either condoned by the state or due to the state's incapacity to protect returning soldiers. Obviously, these are cases where the RPD and Courts have to look beyond what is provided by law.[85]

---

claimant, that deserters publicly opposed to the war were more likely to be prosecuted than other deserters. Her application was subsequently denied and she was deported and transferred to U.S military custody. See 'U.S. War Resister Deported from Canada', *The Globe and Mail*, 20 September 2012, online: http://www.theglobeandmail.com/news/toronto/us-war-resister-deported-from-canada-arrested-at-border/article4556198/ accessed 30 January 2012. See also *Walcott v Canada (Minister of Citizenship and Immigration)*, 2011 FC 415 at paras 37 and 48: 'How, when speaking out about one's political opinion is seen as an aggravating factor by prosecutors in the offence of desertion requiring maximum punishment, does the evidence not support a finding of differential prosecution and therefore persecution on the basis of political opinion?' See also *R.S. v Canada (Minister of Citizenship and Immigration)*, 2012 FC 860, for another case in which a claimant, from Israel, was alleging harsher treatment for selective objectors compared to other army deserters.

83 As argued in *Bakir v Canada (Minister of Citizenship and Immigration)*, 2004 FC 70, para.26.

84 Jones, p. 145. Many cases within the ambit of para.171 will, in any event, involve states where the risk of punishment for deserting would be severe and due process lacking.

85 See for instance *Moslim v Canada (Secretary of State of Canada)* (1994), 75 FTR 243 (TD), p. 244; *Al- Shammari v Canada (Minister of Citizenship and Immigration)*, 2002 FC 364; *Padilla v Canada* (1991), 13 Immigration Law Reporter (2d) 1 (FCA). In a recent case, *Walcott*, para.67, Justice de Montigny examined the risk of hazing in the US military as a punishment outside the framework of 'ordinary laws of general application'.

Is There an Absence of State Protection?

A state is presumed to be able and willing to protect its citizens. To rebut this presumption a claimant needs to provide 'clear and convincing' evidence of a state's inability or unwillingness to protect.[86] The more democratic the state of the claimant is, the more difficult it will be to meet this burden.[87] In *Ward* the persecutor was a non-state actor, and it made sense to require the claimant to demonstrate that state protection was lacking. However, in cases involving selective objectors, the alleged persecutor is the state itself. It was argued in *Hinzman* that, in such a situation, the claimant does not have to demonstrate that the state is at the same time persecutor and unable or unwilling to act as a protector.[88] In that case, the Federal Court of Appeal was invited to consider whether Justice Mactavish's interpretation regarding the application of paragraph 171 to a foot soldier was correct. However, the Court side-stepped that issue by stating that the first issue that had to be addressed was the capacity or willingness of the state to offer protection.

> Where sufficient state protection is available, claimants will be unable to establish that their fear of persecution is objectively well-founded and therefore will not be entitled to refugee status. *It is only where state protection is not available that the court moves to the second stage*, wherein it considers whether the conduct alleged to be persecutory can provide an objective basis for the fear of persecution. If indeed the illegality of the war is relevant, it is at this second stage that the court would consider it.[89]

In *Hinzman* the Federal Court of Appeal did not go past the 'first stage' as it found that state protection was available. The Court noted that the burden on the claimants to rebut the presumption that the United States was capable of protecting them was a heavy one given that this was a democratic country. The claimants 'would be required to prove that they exhausted *all the domestic avenues available* to them without success before claiming refugee status in Canada.'[90] (my emphasis) The

86 *Ward*, p. 724.

87 See *Kadenko v Canada (Minister of Citizenship and Immigration)* (1996), 143 DLR (4th) 532 (FCA) at page 534. In recent cases, the Court has looked at a spectrum among democratic states, noting that the presumption would be easier to reverse in emerging democracies plagued by serious problems such as corruption and drug trafficking. See for instance *Chagoya v Canada (Minister of Citizenship and Immigration)*, 2008 FC 721.

88 See *Hinzman FCA 2007*, para.40. In a decision by the Board that preceded the decision of the FCA in *Hinzman* the RPD has not considered the question of state protection on the basis that 'the alleged agent of persecution on this case is the state itself.' See *Key*, para.10.

89 *Hinzman FCA 2007*, para.42 (my emphasis).

90 *Hinzman FCA 2007*, para.46. Apart from legal remedies within the military justice system, the RPD will also look at whether alternate service is available. If so, and the

Court concluded that the United States had a sophisticated military justice system to deal with these issues and that Hinzman did not make sufficient efforts to have recourse to it.[91]

The Court applied the presumption of state protection even if the alleged persecutor here was the state, contrary to the situation in *Ward* and *Kadenko*. It justified its approach as follows:

> The concepts of persecution and state protection are interconnected such that the question of whether the refugee claimant has attempted to avail himself of the protective mechanisms provided by the state is relevant both where the alleged persecutor is an organ of the state and where the alleged persecutor is a non-state entity.[92]

As a result the Court never answered the argument that Hinzman would be within the ambit of paragraph 171 if the military engagement of the United States in Iraq was illegal and he was punished for opposing it. Hinzman had argued that 'any' punishment in this kind of scenario would amount to persecution. By making a claim for refugee protection conditional upon the exhaustion of all available remedies, the Federal Court of Appeal is exposing potential claimants to whatever punishment their domestic military judicial system will impose on them. This is despite the fact that the Court has not provided any indications as to whether any or some punishment would be contrary to the relevant paragraphs in the *Handbook*.[93] The Court seems to have anticipated in part this argument by noting later that 'approximately 94% of deserters from the US Army have not faced prosecution and imprisonment, but have merely been dealt with administratively by being released from the military with a less-than-honourable discharge.'[94] It added that a soldier would have better chances of getting an administrative discharge if he

claimant does not have a compelling reason not to have applied for it, his or her claim will be rejected. See for instance *Budaghyan*, paras 11 and 13; *Gebre-Hiwet*.

91 Subsequent applications for Pre Removal Risk Assessment and Humanitarian Considerations were also rejected. See *Hinzman v Canada (Minister of Citizenship and Immigration* 2009 CF 415 [*Hinzman 2009*]. However, the Federal Court of Appeal quashed the decision of the immigration officer with regard to the application for Humanitarian Considerations and referred the matter back for redetermination by a new officer. See *Hinzman v Canada (Minister of Citizenship and Immigration)*, 2012 FCA 177 [*Hinzman FCA 2012*].

92 *Hinzman 2007*, para.54.

93 The claimants had argued that they had no chance of succeeding within the military justice system as this system did not protect those who oppose only specific wars and also did not allow challenges to the legality of the war as this was precluded by the US doctrine of 'political questions'. See *Hinzman 2007*, para.55 The Court seems to suggest that Hinzman could have been successful as an absolute conscientious objector. See para.58. However, that claim had been rejected at least once by the US military as the Court noted early in its decision. See para.12.

94 *Hinzman 2007*, para.58.

went through the system instead of deserting. However, even if one accepts these conclusions,[95] it still does not provide a satisfactory answer to the argument put forward by the claimant and not addressed by the Federal Court of Appeal that in some cases 'any' punishment would amount to persecution. A better approach would have been to review Justice Mactavish's decision to clarify the law on the various issues that still need to be addressed.[96]

The emphasis on the demonstration of 'absence of protection' as a first stage is a shift in cases dealing with selective objectors. While in the past it was barely considered,[97] now it is a key issue. Not surprisingly, since that decision, many cases have faltered at the first stage of the analysis.[98]

However, in a recent case, *Vassey*, the claimant directly challenged the issue of state protection in the US and was successful in his application for judicial

95 As the Court noted, *Hinzman 2007*, para.59, the counsel for the claimant challenged these statistics. He argued that these reflected the situation before the most recent military operations by the US. The Court mentioned that they were not made aware of contrary evidence and that, in any event, there were reasons to believe that these statistics would not change materially. However, for contrary evidence see fn.74.

96 One would have thought that the Court would have liked to reconcile its previous decision in *Al-Maisri* with the judgment of Mactavish in this case, as this judge invited the Federal Court of Appeal to do. In *Lebedev*, para.24, Justice de Montigny noted the opportunity missed by the Federal Court of Appeal in *Hinzman 2006*: 'As a result, there is still no definitive pronouncement on how to properly interpret paragraph 171 of the UNHCR Handbook – and particularly, whether the unlawfulness of a given conflict is relevant to the refugee claim of an ordinary foot soldier.' See also Edward C. Corrigan, 'American War Resister Cases in the Canadian Federal Court' (2009) 81 Immigration Law Reporter (3d) 233, p. 242.

97 See the comment of Martin Jones prior to *Hinzman 2006,* in Martin Jones, 'Beyond Conscientious Objection: Canadian Refugee Jurisprudence on Military Service Evasion' (CRS Working Paper Series, 2005), p. 6: 'State protection is seldom discussed in the context of military service evasion.' He notes, p. 7, that it would come up mostly when the claimant feared forced conscription by guerrilla groups.

98 A good example is the decision in *Key*. In its decision in 2008 the Federal Court returned the matter to the RPD as it had not initially considered the issue of state protection on the basis that the state was the persecutor. In a subsequent decision, the RPD rejected the claim on the basis of adequate state protection in the United States. Leave for judicial review was denied. See *Key v Canada (Minister of Citizenship and Immigration),* 2011 FC 92 [*Key 2011*]; see also *Colby*; *Landry v Canada (Minister of Citizenship and Immigration),* 2009 FC 594. In the *Landry* case the claimant did exercise some remedies available within the US military system but deserted when he was told that, instead of the discharge he was expecting, he would likely get 30 days in 'correctional custody' and then he would be sent to join his unit in Iraq. The Court agreed with the RPD that the claimant had not exhausted all the legal remedies available to him. The RPD, para.23, had noted that the claimant 'still had legal avenues open to him, including court-martial process and other legal steps up to and including *certiorari* to the U.S. Supreme Court.' A number of US selective objectors whose claim for refugee protection were rejected subsequently applied for permanent residence under humanitarian considerations. See for instance *Hinzman FCA 2012* (application returned for new determination); *Lowell* (application denied).

review to set aside the rejection of his refugee claim by the RPD. The claimant had been deployed in Afghanistan. He testified that he was ordered to perform actions contrary to the rules of armed conflict, including the following:

> These orders included raiding civilian homes and recognizance by fire where his unit pre-emptively fired on a location where they believed the enemy forces were located without taking any precautions to ensure that civilians were not harmed. The applicant further stated that he was part of the supervision of the Afghan National Army which he learned were placing detainees in 'hot boxes' under extreme conditions to obtain information. Finally, the applicant alleges that his unit strapped the dead bodies of Afghan insurgents to US military vehicles and drove through villages in order to intimidate local populations.[99]

He claimed refugee protection without having first resorted to the US military judicial system. A number of experts retained by the claimant explained that recourse in the US would have been useless for a number of reasons:

- The existing law in the US would have prevented him from invoking, as a defence to a charge of desertion, the fact that he was asked to follow unlawful orders.
- There was no state protection against the discriminatory exercise of prosecutorial discretion by the commanding officer, directed mostly against those deserters publicly opposed to the wars the US was engaged in. Not only was the state more readily prosecuting vocal selective objectors, but their public opposition also constituted an 'aggravating factor' in sentencing.
- The US military judicial system, by Canadian standards, was lacking in terms of security of tenure, financial security and institutional independence. In addition, complaints of unjustified command influence often proved to be ineffective.

In setting aside the RPD's decision, the Federal Court noted that the RPD did not explain how and why it reached its conclusions with regard to the availability of state protection in spite of all the expert evidence to the contrary. Of course, absence of protection is only the 'first stage'. The claimant still has to demonstrate that either he falls within the parameters of paragraph 171 or, as he also alleges here, that his punishment, contrary to paragraph 169, is discriminatory on the basis of one of the Convention's enumerated grounds.

When a complaint is not accepted at the lower level of enforcement, for instance a complaint to a police officer, the claimant will be expected to go further and complain to a superior officer.[100] However, the assessment of a state's willingness to protect should also take into account the context particular to the army. In the

---

99 *Vassey v Canada (Minister of Citizenship and Immigration)*, 2011 FC 899, para.4

100 See Kadenko.

Smith decision, the claimant had complained to her direct superior and was told by the latter not to complain further up the chain of command. Justice de Montigny, without opining whether the claimant should have complained to higher rank officers, observed that: 'one must take into account the particular environment that an applicant finds himself in. It is clear that in the Army reigns an atmosphere of unconditional obedience to the hierarchy.'[101]

Conclusion

Beginning with the Federal Court of Appeal decision in *Zolfagharkhani*, Canadian courts have identified key legal questions involved in claims by selective objectors and have resolved some of them. As a result, ordinary laws of general application no longer constitute an absolute bar to these claims. As for the interpretation of paragraph 171 of the *Handbook*, some Federal Court judges have adopted a more liberal interpretation. In their view, selective objectors do not have to establish that their conduct on the ground would have amounted to an excludable action. As well, these judges have stated that international condemnation is not a prerequisite to a claim for refugee protection. On the other hand, the primacy given to the assessment of state protection has resulted in the denial of many claims and has precluded the Federal Court of Appeal, at least until now, from venturing into further analysis of important issues. However, in closing, one should put the hurdle posed by 'state protection' into perspective. While it will prove difficult for a claimant from a country with a long democratic tradition to go beyond this first stage, democracy is unfortunately not well rooted or is absent in many countries throughout the world. Deserters or draft evaders from these countries should be able to meet the burden of establishing the absence of state protection. These cases will provide the opportunity for the RPD and the courts to provide more in-depth analysis regarding selective objectors. In addition, they may prompt the Federal Court of Appeal to revisit and widen the avenues for claims by absolute conscientious objectors, an issue that, in the words of Justice de Montigny, needs to be revisited:

> Until the Federal Court of Appeal provides further clarification, I feel bound to follow its most recent decision on the subject in *Ates*, above. However, in my view, the issue of conscientious objection still raises a host of outstanding questions, begging for resolution.[102]

---

101 *Smith v Canada (Minister of Citizenship and Immigration)*, 2009 FC 1194, para.43. See a subsequent decision, *Smith v Canada (Minister of Citizenship and Immigration)*, 2012 FC 1283 where the Federal Court upheld the decision of the Refugee Protection Division to reject the claim on the basis of lack of credibility and availability of state protection.

102 *Lebedev*, para.50.

Chapter 11
Conscience in Lieu of Obedience: Cases of Selective Conscientious Objection in the German Bundeswehr

Jürgen Rose

Freedom of Conscience and the Right to Conscientious Objection in Germany

After World War II and the fall of the Third Reich, the new West German constitution, the so-called 'Basic Law for the Federal Republic of Germany', was adopted and proclaimed in May 1949. In it every citizen was guaranteed the fundamental human right to freedom of faith and conscience as well as the right to refuse to perform military service involving the use of arms against their conscience. The three paragraphs of the relevant Article 4 read as follows:

> (1) Freedom of faith and of conscience, and freedom to profess a religious or philosophical creed, shall be inviolable.
> (2) The undisturbed practice of religion shall be guaranteed.
> (3) No person shall be compelled against their conscience to render military service involving the use of arms. Details shall be regulated by a federal law.

While paragraph 1 caused little discussion, paragraph 3 evoked fierce debates, because some, such as the later Federal President Theodor Heuss, suspected that, 'if we put here simply conscience, we will enshrine in the constitution a "mass over-use of conscience" in the event of war'.[1]

However, in the end he could not prevail against the stand taken by Social Democrats like Carlo Schmid, who replied that '[t]he citizen must be enabled to state for moral reasons: "I wish to serve my fatherland in its crisis in a manner other than by killing someone"',[2] and Fritz Eberhard, who argued that

> we have gone through a 'mass slumber of conscience'. During that 'mass slumber of conscience' millions of Germans said: orders are orders, and killed in view of

1 Quoted in Gerd Bucerius, 'Verweigerung und Gewissen. Das Grundgesetz wurde gegen seine Väter ausgelegt', *Die Zeit*, 23 December 1977 http://www.zeit.de/1977/52/verweigerung-und-gewissen all web links accessed 31 January 2013.

2 Ibid.

> that. That paragraph can have great pedagogical effect, and we hope it will do … this is why I believe that, precisely in this situation after the war and after the totalitarian system, now that we want to break with this notion that orders are orders – when in fact we want to build democracy – that paragraph is apposite.[3]

One may wonder why the issue of conscientious objection was discussed as early as in January 1949, when nobody was talking about rearming Germany or even creating a new German Army. However, there were several driving factors behind the debate. Most urgent was certainly the fear that during the emerging Cold War the occupation powers – adamant that the *Krauts* should defend their own soil instead of shedding the blood of allied soldiers if the conflict turned into a hot war – would try to compel Germans to serve in some kind of military formations commanded by foreign generals of the Allied Forces. The second and very important factor was the proximity of the experience that military courts during the Third Reich[4] had systematically sentenced unprecedented numbers of conscientious objectors to death and executed them according to the maxim: 'He who fights may die, he who refuses to fight must die!' Last but not least, the third consideration was that in the late 1940s conscientious objection was not yet internationally accepted as a fundamental human right. That is why the basic right to universal conscientious objection, but not to selective conscientious objection, was embodied in the new German Basic Law so early. As a result everybody, including soldiers no matter whether they are active, in reserve or retired, and regardless of whether they are conscripts or volunteers, may make use of that right at any time.[5]

While universal conscientious objectors, after many years of fierce controversy among supporters and opponents of the new German Bundeswehr, were increasingly accepted by the broad German public, this was not the case for those who additionally claimed a right to selective conscientious objection. It was only in 2005 that the Federal Administrative Court provided explicit clarification on this matter with its judgement in a lawsuit against a Major in the German Army who had refused to obey several orders during the war against Iraq on moral grounds.[6] First of all, the judges explained unmistakably that Article 4 paragraph 1 of the German Basic Law unreservedly applies to soldiers, too. Secondly, they conceded that soldiers may experience severe moral conflict by obeying an order

3 Quoted in Helmut Donat, Günter Knebel, Karl-Ludwig Sommer and Peter Tobiassen, 'Der Deutsche Ausschuss für Fragen der Wehrdienstverweigerung 1953-1957', *Zentralstelle KVD*, undated, http://www.zentralstelle-kdv.de/z.php?ID=116

4 On this point, see Manfred Messerschmidt, *Die Wehrmachtjustiz 1933–1945* (Paderborn: Ferdinand Schöningh, 2005).

5 However, for a professional soldier such an endeavour would not be easy at all, because he is legally obliged to meticulously expose his more or less sudden change of conscience to a particular board of examiners that is in charge of granting his motion for conscientious objection.

6 The case of Major Florian Pfaff is described below.

that clashes with their conscience. Under such circumstances, in their opinion, one must not expect them to obey the order, and hence they are entitled to reject it on the basis of the individuals' constitutional right to the inviolability of conscience. However, the subject of this lawsuit involved not only the right to refuse a single specific order, but also the question of selective conscientious objection. That is, because the Major had argued that obeying his orders would have involved him in a war of aggression which he considered completely and utterly impossible not only on personal moral grounds, but also because of the German Constitution and international law. Hence perhaps the most important feature of the case was that an active field officer resolved upon an act of selective conscientious objection and the Federal Administrative Court, subordinate only to the highest German court, the Federal Constitutional Court, accepted such a courageous stance for the first time in the history of post-war Germany.

Of inestimable significance within the context of the lawsuit described above was that the judges explicitly stressed the unconditional commitment of the armed forces to law and order: 'The Basic Law imposes the commitment of the armed forces to basic rights, but does not restrict basic rights on decisions and requirements of the armed forces.'[7] That judgement implies not only that the military is committed to the Constitution, but also that all of its members are strictly obliged to follow the special military laws and regulations, for example the 'Soldatengesetz' (Law on Soldiers) and the 'Wehrstrafgesetz' (Law on Military Jurisdiction) of the Federal Republic of Germany. The relevant articles of paragraph 10 of the Law on Soldiers on the 'duties of the superior' state that:

> (4) He is authorized to issue orders for official reasons only, and only when observing the rules of international law, national law and service regulations.
> (5) He is responsible for the orders he has given. He has to enforce orders by reasonable means.

Paragraph 11 of the Law on Soldiers (§ 11 SG) is also relevant:

> (1) The soldier has to follow his superior. He must observe their orders completely, precisely and immediately. Disobedience does not apply when he refuses to obey an order that violates human dignity or that was not issued for official reasons; the erroneous assumption that such an order is given releases him from his responsibility only if he was not able to avoid his error, and if he could not have been expected in the given circumstances to defend himself against the order by remedy.

7 Bundesverwaltungsgericht, *Urteil des 2. Wehrdienstsenatsvom 21 Juni 2005 – BVerwG 2 WD 12.04*, p. 112, http://www.bundesverwaltungsgericht.de/media/archive/3059.pdf.

(2) An order must not be observed if that implies a criminal offence. If the subordinate follows the order anyhow, he will become guilty only if he recognizes, or if it was obvious for him in the given circumstances, that obeying implied a criminal offence.[8]

In summary, the evolution of the modern German armed forces after World War II, as well as the recent jurisdiction on military affairs, reveals the significance of two essential aspects of what might be called, in short, the German 'military constitution'. These are the strong commitment of individual soldiers to their conscience as well as the strong commitment to law and order. These two elements also constitute the foundations of the revolutionary new concept of 'Inner Leadership' in the German Bundeswehr.

The Philosophy of *Innere Führung* (Inner Leadership)

Kant defines a state as a 'union of men under law'. By extension such a state's army must then be considered an assembly of free, republican citizens under arms for the purpose of protecting that very polity. It is on the basis of this logic that Wolf Graf von Baudissin, three-star General military philosopher, peace researcher and 'father' of Innere Führung, made the 'citizen in uniform' the centre of his concept or philosophy.[9] That is because Innere Führung calls for situating such a 'citizens' army' within the framework of a democratic, according to Kant, republican, state under the rule of law that is also embedded in an international peace order (Friedensordnung) as a subject of international law. The historical background and the starting point for military reform based on the concept of Innere Führung was the catastrophe of the Second World War during which the traditional German military, that is the Wehrmacht, had willingly served the National Socialist dictatorship in a war of aggression and genocide. It, or at least its senior leadership, was utterly delegitimized at the end of the War. Rebuilding the German armed

8 Bundesministerium der Verteidigung Führungsstab der Streitkräfte InfoM (ed.): *Grundgesetz für die Bundesrepublik Deutschland. Werte und Normen für Soldaten* (Bonn, 2003), p. 112f. The Law on Military Jurisdiction constitutes the basis for prosecuting offenses against the legal norms mentioned before. The most relevant are: section 5 (acting under orders), section 19 (disobedience), section 20 (refusal to obey orders), section 21 (careless failure to follow an order), section 22 (bindingness of an order, error), section 32 (misuse of command authority for illegal objectives), section 33 (incitement to an illegal action), section 34 (unsuccessful incitement to an illegal action). See *Wehrstrafgesetz WStG*, March 30, 1957, BGBl I 1957, 298, revised by proclamation of May 24, 1974 I 1213, http://www.juris.de

9 Wolf Graf von Baudissin, 'Referat auf einer Tagung für ehemalige Soldaten in der Evangelischen Akademie Hermannsburg am 3. Dezember 1951', in Klaus von Schubert (ed.), *Sicherheitspolitik der Bundesrepublik Deutschland. Dokumentation 1945–1977, Teil II* (Bonn: Bundeszentrale für Politische Bildung, 1978), p. 356.

forces and integrating them into the Atlantic Alliance was driven by a desire for a complete break with the past – never again should the German military commit crimes against international law. With this in mind, Innere Führung amounts to a code of the armed forces of the Federal Republic of Germany; it embodies so to speak the basic legal and ethical underpinnings of the Bundeswehr.

The answer General von Baudissin once gave to the fundamental question of how to define Innere Führung was: 'demilitarization of the military's self-conception.'[10] That precept refers to three dimensions of the military profession, namely an organizational one, a societal one, and an international one. All three of them ultimately merge in one focal point that can, positively put be defined as *civilizing the military.*[11] That goal of civilizing the military is achieved when service in the armed forces is compatible with human rights, democracy and peace, and when, as Graf von Baudissin once posed, 'democracy does not end at the barrack's gate.'[12]

Considering first of all the military, Innere Führung safeguards the fundamental human and civil rights the constitution guarantees to all citizens, including when they are doing their military duty in the Bundeswehr, because those rights are to be defended in case of emergency (Ernstfall), if necessary by risking life and limb. Considering the functional imperatives that apply within the military as a 'total' institution, this idea seems bold or even revolutionary. Innere Führung strives to overcome the suppression of human individuality that prevails within a tightly regulated system rooted in strictly enforced order and obedience. At the same time, the role model of the *citizen in uniform* who is critically-minded, discriminating and has the courage of his or her convictions, is intended to banish a spirit of subservience once and for all from the military:

> There can no longer be a relationship between subordinates and superiors, or a concept of the military as a 'tool of combat' among citizens who are united to perform their duty under arms, but here they are clearly partners in different functions with equal dignity following equal responsibility.[13]

Secondly, Innere Führung defines a fundamentally new relationship between the military and society. Historically, the German military had been shaped by an elitist, closed shop mentality, the so-called esprit de corps that prepared the ground for the disastrous notion of a state within a state. The revolutionary essence of

10 Wolf Graf von Baudissin and Günter Will, 'Die schwere Geburt. Zur Entwicklungsgeschichte von Truppeninformation', in *Information für die Truppe*, 8 (1991), p. 64.

11 Detlef Bald, *Die Bundeswehr. Eine kritische Geschichte 1955 – 2005* (München: C. H. Beck, 2005), p. 32.

12 Wolf Graf von Baudissin, 'Redeanläßlich der Verleihung des Freiherr-vom-Stein-Preises am 10. Februar 1965 in Hamburg', in Schubert, *Sicherheitspolitik*, p. 412.

13 Wolf Graf von Baudissin, *Soldat für den Frieden* (München; Piper, 1969), p. 207.

General von Baudissin's concept for military reform was first of all marked by the idea that the armed forces should be made fit for democracy and compatible with a pluralistic society by overcoming the traditional narrow-mindedness of military thinking. This was to prevent the self-isolation of the armed forces as well as their isolation from society, and to promote both their integration into the democratically structured state and their conformity with an open, pluralistic society.

Thirdly and finally, Graf von Baudissin's rationale for the structure and organization of the new German army sought from the start to avoid any reduction to a solely national dimension. He was determined to set up the Bundeswehr within the framework of a European security architecture – that is, tied to an international perspective.[14] Innere Führung is based on the fundamental assumption that in the nuclear age the soldier must first and foremost work towards the preservation of peace; no longer is the battlefield the place of proving him or herself, and 'during peacetime the question of combat motivation is not under consideration.'[15] War can no longer be seen as a normal means of politics. Ultimately, what matters is the preservation of existence. Thinking in categories of war fighting capabilities is obsolete, what is essential is the military's suitability for peace. Given the new scope of risks since the end of the Cold War, there is no getting around that insight.[16] Striving for military victory over international terrorism and stemming the proliferation of weapons of mass destruction with strategies of preventive warfare is a deadly illusion. This is all the more true in view of the exceedingly controversial nature of recent wars in which the US and its allies have been engaged.

Old Values – New Missions

In times of the *Global War on Terror*, preventive warfare, crimes against international law, torture, and threats to fundamental human and civil rights, the frequently inhumane reality of the use of military force may create the suspicion that a soldier is just a kind of battle robot who follows orders without thinking.[17] This perception is enhanced when even NATO allies, all Western democracies, seem to abuse their

14 Wolf Graf von Baudissin, 'Innere Führung und Sicherheitspolitik', in *Information für die Truppe*, 8 (1987), p. 23.

15 Wolf Graf von Baudissin: *Staatsbürger in Uniform*, in Bundesministerium der Verteidigung Fü S I 3 (ed.), *Legitimation soldatischen Dienens*, Schriftenreihe Innere Führung, supplement No. 1/87 to Information für die Truppe (Bonn 1987), p. 98.

16 Dieter S. Lutz, 'Graf Baudissin – Reformer, Wissenschaftler, Hochschullehrer. Zum Gedenken an den Gründungsdirektor des IFSH', in Hilmar Linnenkamp and Dieter S. Lutz (eds.), *Innere Führung. Zum Gedenken an Wolf Graf von Baudissin* (Baden-Baden; Nomos, 1995), p. 14.

17 Jürgen Rose, 'Gehorsam bis zum Hindukusch? Das Grundgesetz verbietet die Vorbereitung von Angriffskriegen. Trotzdem wird ein Bundeswehrsoldat, der sich weigert mitzumachen, degradiert', *Sicherheit und Frieden – Security and Peace*, 22/3 (2004), p. 134 and Rose, 'Gehorsam bis zum Verfassungsbruch. Der 20. Juli und die Pflicht zur

armed forces in missions whose legality under international law is disputed.[18] Not least because of this, civil society has become aware of the problem of legitimacy of such wars under international and constitutional law. The number of cases of disobedience in many intervention and occupation forces reveals that even amongst military professionals, who are expected to execute the combat missions agreed on by the political leadership, sensitivity has grown regarding the implications of the legal constraints on the initiation and conduct of war, including the Rome Statute of the International Criminal Court and the implications this has for both for the legal and the moral dimensions of military service.

The fundamental question service members who operate within the area of tension between their duty to obey, loyalty to the law, and freedom of conscience must answer is: How may or shall or must I act as a soldier who is strictly subject to the primacy of politics, if my political leaders or military superiors urge me into a war that will inevitably bring death and injury to human beings, if this war is potentially or even clearly a war of aggression, which clearly constitutes a crime against international law?

The service member involved has no choice but to confront the problematic nature of that vitally important issue. Since the Nuremberg Tribunals, service members can no longer appeal to a superior political or military authority in order to avoid responsibility for obeying illegal orders. In accordance with Kant's philosophy of law and morality,[19] the reason for this is the recognition that the conscience of each person sets the standard for any human action. For the service member this means that superior orders alone are insufficient to legitimize any military action. By obeying an order soldiers voluntarily transform the extrinsic will of another person into their own intrinsic determination, and before actualizing their own intention through their action soldiers have to scrutinize its legitimacy in relation to their conscience.[20]

Modern philosophy of law founded upon the traditions of the Enlightenment is reflected in the so-called Nuremberg Principles.[21] On the one hand the latter

---

Verweigerung. Wer sich in der Bundeswehrdaranhält, kann degradiert werden, wie der Fall des Majors Florian Pfaff zeigt', *Freitag – Die Ost-West-Wochenzeitung*, 23 July 2004, p. 4.

18 On this point in detail, see Jürgen Rose, *Ernstfall Angriffskrieg. Frieden schaffen mit aller Gewalt?* (Hannover: Ossietzky, 2009).

19 Immanuel Kant, *Grundlegung zur Metaphysik der Sitten* (Hamburg: Felix Meiner, 1965).

20 This implies that ultimately every soldier decides with which means and up to which point he would be prepared to fight. The functional imperatives of the military subsystem, at least in its current constitution, are incompatible with this moral imperative, which suggests, to paraphrase Hegel – this is so much the worse for the military.

21 On this point, Gregor Schirmer (ed.), *Die Nürnberger Prinzipien – ein Umbruch im Völkerrecht*, Bulletin für Faschismus- und Weltkriegsforschung (Academic Half-Year Journal, ed. in cooperation with the Berliner Gesellschaft für Faschismus- und Weltkriegsforschung e. V.) (Berlin; Edition Organon, 2006), and; Herbert R. Reginbogin (ed.) *Die Nürnberger Prozesse: Völkerstrafrecht seit 1945* (München: De Gruyter, 2006).

impinged on various national bodies of military law.[22] On the other hand they were affirmed at the level of international law, for instance by the Code of Conduct on Political and Military Aspects of Security, which states in paragraphs 30 and 31 that:

> 30. Each participating State will instruct its armed forces personnel in international humanitarian law, rules, conventions and commitments governing armed conflict and will ensure that such personnel are aware that they are individually accountable under national and international law for their actions.
> 31. The participating States will ensure that armed forces personnel vested with command authority exercise it in accordance with relevant national as well as international law and are made aware that they can be held individually accountable under those laws for the unlawful exercise of such authority and that orders contrary to national and international law must not be given. The responsibility of superiors does not exempt subordinates from any of their individual responsibilities.[23]

Time after time high-ranking military leaders have acknowledged and corroborated the axiom of law referring to the personal responsibility for one's activities, which applies to each soldier no matter whether superior or subordinate. Major General Hans Peter von Kirchbach commented on this in 1992:

> The tension between liberty and obedience is manifest in [a soldier's] obedience to carry out orders on the one hand, and in obedience to a value system on the other. The tension is between obedience and allegiance to the state on the one hand and the knowledge that national action is the next-to-last possibility. It means that a conscience bound to a higher system of values is the decisive court of appeal. Certainly, the state will not expect its citizens to act against the advice their conscience gives them. But an awareness of this tension and the realization that not every expectation can be fulfilled makes the ultimate difference between the soldier and the mercenary.[24]

---

22 The relevant regulations of the German military law are discussed above. In the United Kingdom soldiers are obliged to refuse any illegal order. In Denmark and France soldiers must refuse all obviously illegal orders. In addition, they are authorized to refuse all other illegal orders. In Belgium, Luxembourg, the Netherlands, Poland, and Spain, soldiers must refuse to obey illegal orders. But while a soldier in the Netherlands may refuse to obey any illegal order, soldiers in Germany, Luxembourg, and Spain are only authorized to refuse to obey a limited range of illegal orders. These embrace primarily orders that hamper human dignity; on this point, see Georg Nolte and Heike Krieger, *Europäische Wehrrechtssysteme* (Göttingen: Nomos, 2002), and Jürgen Groß, *Demokratische Streitkräfte* (Baden-Baden: Nomos, 2005), pp. 93f.

23 Auswärtiges Amt (ed.), *Von der KSZE zur OSZE. Grundlagen, Dokumente und Texte zum deutschen Beitrag 1993-1997* (Bonn: 1998), pp. 267f.

24 Hans Peter von Kirchbach, 'Offizier im Heer der Einheit', in *Truppenpraxis*, 4 (1992), p. 335.

Two years later former Chief of Staff of the German Armed Forces, four-star General Klaus Naumann, even posited a soldier's obligation to insubordination in his official report to the General Inspection Committee, when he stated:

> According to our conception of the rule of law and ethics, the commander's claim to obedience is faced by both the right and the obligation to insubordination, where this rule of law and morality is no longer in accordance with the military mission. The soldier therefore would be positioned outside the free and democratic legal order.[25]

Also, former US Secretary of Justice Ramsey Clark once said about the limits of the soldier's duty to obey orders: 'The greatest cowardice consists in obeying an order which requires a morally unjustifiable action.'[26]

Selective Conscientious Objection and Situational Refusal to Obey Orders in the German Armed Forces

Those considerations, regularly invoked by the Bundeswehr, were on the one hand posed with regard to the classic mission of defence of both the nation and the Alliance, and on the other hand referred back directly to the tradition of the military resistance fighters who attempted to assassinate Hitler on 20 July1944. In their original historical context they would have appeared merely theoretical and abstract. But if the exercise of command and the execution of orders is strictly bound by both domestic and international law, this could not a priori exclude soldiers from occasionally refusing to take part in military missions which were either obviously illegal, doubtful or controversial under international law. And indeed, after the end of the Cold War, this matter became unexpectedly explosive, especially since the Bundeswehr has claimed to be defending Germany in the Balkans or even the Hindu Kush. Several members of the German armed forces preferred to follow their conscience as well as their oath of duty instead of obeying the orders of their superiors which they considered to be irreconcilable with constitutional norms and international law.[27] All those conscientious objectors were acting under severe personal risk – after all insubordination and

25 Klaus Naumann, *Generalinspekteursbrief 1* (Bonn 1994).

26 Ramsey Clark, *Wüstensturm. US-Kriegsverbrechen am Golf* (Göttingen: LamuvVerlag, 1993), p. 268.

27 Jürgen Rose, *Gewissensnöte – Bundeswehr-Soldaten, die sich Auslandseinsätzen verweigern*. script for NDR Info *Das Forum*, 'Streitkräfte und Strategien' by Andreas Flocken, July 10 (2004), pp. 11-15 and Rose, 'Und sie verweigern doch! Mehr als ein Einziger. Seit die Bundeswehr an Interventionseinsätzen teilnimmt, sind bereits mehrfach Soldaten ihrem Gewissen gefolgt – Replik zu Andreas Zumach im "Freitag" 02/08', *Freitag – Die Ost-West-Wochenzeitung*, 1 February, 2008, p. 4.

disobedience are criminal offenses under penalty of imprisonment under the German military penal code.[28]

So as early as 1999, in the course of NATO's air war against the Federal Republic of Yugoslavia, around a dozen German Air Force fighter pilots refused to board their ECR-TORNADO aircraft and fly the assigned SEAD [Suppression of Enemy Air Defences] missions.[29] At the time this event attracted hardly any media coverage, presumably mainly because the government shied away from a trial which might be fought all the way up to the Federal Constitutional Court or the European Court of Justice and would have attracted considerable media coverage. However, subsequently the German armed forces' leadership was to show far less caution in dealing with the insubordinate behaviour of soldiers, as some significant cases illustrate. They are particularly relevant for both their juridical assessment and the sanctions they entailed. They will be considered in greater detail below. The first case concerns that of Major Florian Pfaff, the second the author himself, and the third medical orderly Sergeant Major Christiane Ernst-Zettl. In addition, there are several confirmed cases of conscientious objection by active or retired soldiers including officers, some of them accepted and others not, in which the protesters explicitly refer to the missions of the Bundeswehr as justification for their decisions of conscience.

Major Florian Pfaff

As far as is known, Major Florian Pfaff[30] was the only soldier among the entire German armed forces who refused to obey orders which, had he carried them out, would have led him to knowingly participate in what he considered a war of aggression launched by the US and Britain and which the well-known German law philosopher Reinhard Merkel had previously denounced as an 'outrage against

28 This does not imply any value judgement about the empirically very manifold manners of conscientious objection that range from AWOL, permanent desertion, or situative refusal to obey orders, to conscientious objection for principled reasons. Referring to the typology of objectors: Anonymous, *War Heroes of the Iraq War – War Resisters from within the Military* http://www.tomjoad.org/WarHeroes.htm and; UnitedforPeace.org (ed.), *Iraq War Resisters*, http://www.unitedforpeace.org/article.php?id=3455

29 Jürgen Rose, *Gewissensnöte*, p. 13, and; *Security and Peace, 3* (2004), p. 135.

30 For the case of Florian Pfaff see inter alia Hans Georg Bachmann, *Militärischer Gehorsam und Gewissensfreiheit*, in Holger Zetsche and Stephan Weber (eds), *Recht und Militär. 50 Jahre Rechtspflege der Bundeswehr* (Baden-Baden: Nomos, 2006), pp. 156-68; Manuel Ladiges, 'Irakkonflikt und Gewissenskonflikte', *Wissenschaft & Sicherheit* online 02 (2007), March 22, 2007, www.sicherheitspolitik.de (there one can find further references to commentaries of that judgement); Jürgen Rose, 'Gewissensfreiheit statt Kadavergehorsam. Freispruch für Bundeswehroffizier', in *Wissenschaft und Frieden*, 1 (2006), pp. 44-7.

international law'.[31]Actually Major Pfaff was requested to contribute to a software program which he and the superior whom he consulted on that issue could not rule out from being used directly or indirectly to give logistic and technical support to the war against Iraq. After combat activities in Iraq had started, Major Pfaff told his superiors that he would not obey any orders if by executing them he would become guilty of collaborating in the 'murderous occupation of Iraq by the US (and others)'.[32]

Immediately, disciplinary action was taken against the Major in April 2003, in the course of which he was demoted to the rank of captain with a court martial in February 2004.[33] Both the prosecutor and Pfaff himself lodged an appeal to the Federal Administrative Court against this original judgment. Almost one and a half years later the Second Senate for military jurisdiction of the Federal Administrative Court rescinded the decision of the initial court martial, rejected the appeal of the prosecutor as ill-founded and – in a judgement described as spectacular[34] – granted Major Pfaff a full discharge from one of the most serious charges that could be raised against a service member: insubordination and disobedience.

With their judgment the judges considerably enlarged the scope of discretion regarding that issue for each soldier, to cover even cases of uncertainty concerning the legitimacy of a military intervention.[35]A German soldier who is confronted with a moral conflict and is able to explain it in a serious and credible manner need not obey orders if, by executing them, he or she would be involved in legally 'grey area' activities. With its decision the Federal Administrative Court de facto reassigned the burden of proof. It is no longer the soldier who has to prove that his or her refusal to follow orders was required by law, but the government that must explain to the 'citizens in uniform' sent into battle that their mission complies with both international and constitutional law.[36] For the Bundeswehr, an army under

---

31 Reinhard Merkel, 'Krieg. Was Amerikaaufs Spiel setzt. Ein Präventivkrieg mag der Logik imperialer Macht entsprechen. Aber er untergräbt das Rechtsbewußtsein der Menschheit', in Kai Ambos and Jörg Arnold (eds), *Der Irak-Krieg und das Völkerrecht* (Berlin: Berliner Wissenschafts-Verlag, 2004), p. 28. See also, Dietrich Murswiek, *Die amerikanische Präventivkriegsstrategie und das Völkerrecht*, ibid. See also the judgement from 21 June 2005 of the Second Senate for military jurisdiction of the Federal Administration Court BVerwG 2 WD 12.04, http://www.bundesverwaltungsgericht.de/media/archive/3059.pdf

32 Florian Pfaff quoted from Bundesverwaltungsgericht: loc. cit., p. 103.

33 Truppendienstgericht Nord, *Urteil im gerichtlichen Disziplinarverfahren – N 1 VL VL 24/03 – vom 9. Februar 2004* (unpublished). Although the First Chamber took for granted that the attack on Iraq had been a war of aggression (!), the judges denied the soldier's right to refuse his orders because they did not accept the soldier's personal involvement as cogently proven. Therefore they found him guilty of breaches of article 7 (obligation for devote service), article 10 (duties as superior), article 11 (obligation to obey orders) and article 17 (obligation to preserve respect and trust) of the German Military Code.

34 Bundesverwaltungsgericht, loc. cit.

35 Bundesverwaltungsgericht, loc. cit. p. 72.

36 Bundesverwaltungsgericht, loc. cit. p. 116.

the rule of parliament, the implications of the Federal Administrative Court's judgement are of vital relevance. This is, because the primacy of politics applies only within the limits of law and statutes; beyond these the primacy of conscience rules. Otherwise why did the Federal Administrative Court state: 'In case of a conflict between conscience and legal obligation the freedom of conscience is "inviolable".'[37]

Lieutenant Colonel Jürgen Rose

In order to play down its relevance as an unusual, isolated case, the German military leadership successfully prevented the Pfaff decision from becoming widely known in public,[38] and, the litmus test on that historic judgment had to wait till 15 March 2007, when the author of the present text submitted a so-called 'official request' to his disciplinary superior, which read:

> With regard to the decision of the German government to deploy TORNADO weapon systems of the German Air Force to the Afghan war theatre, the German Parliament's approval on 9 March 2007 and the task order issued by the Armed Forces Logistics Command in the meantime, I hereby declare that it goes against my conscience to support the operation of TORNADO weapon systems in Afghanistan, because I cannot exclude the possibility that by virtue of my own action I will contribute to a mission of the German Armed Forces against which there are grave reservations with respect to constitutional, international, criminal and international criminal law. In addition, I request to be exempted from all further tasks connected with 'Operation Enduring Freedom' in general, as well as from the deployment of the TORNADO weapon systems in particular.[39]

The above request above was preceded by an 'official statement' made one year previously which declared inter alia:

> Whilst appreciating the primacy of politics and committed to my oath of duty, which is to devotedly serve the Federal Republic of Germany as well as to bravely

---

37 Bundesverwaltungsgericht, loc. cit. p. 106.

38 On this point, see Jürgen Rose, *Ignoriertes Urteil? Wie die Bundeswehrmit der Leipziger Entscheidung zur Gewissensfreiheit umgeht*, script for NDR Info Das Forum, »Streitkräfte und Strategien« by Andreas Flocken, 14 January (2006). See also: 'Gewissensfreiheit statt Kadavergehorsam. Freispruch für Bundeswehroffizier', in *Wissenschaft und Frieden*, 1 (2006), pp. 44-7.

39 Jürgen Rose, 'Aufklären, damit die anderen bomben können. Dokumentation. Antrag des Oberstleutnants Jürgen Rose, von allen dienstlichen Aufgaben bei einem Tornado-Einsatz in Afghanistan entbunden zu werden', in *Freitag – Die Ost-West-Wochenzeitung*, 23 March 2007, p. 7.

> defend the right and freedom of the German people, I hereby declare that it goes against my conscience to follow any orders which violate international or German law. While doing so I refer to article 4, paragraph 1 of the German constitution as well as to the decision of the Federal Administrative Court regarding freedom of conscience given on June 21, 2005 (BVerwG 2 WD 12.04).[40]

That 'official statement' had been entered in my personal file without objection. While that announcement expressed my conscientious convictions regarding my service as a staff officer of the German Bundeswehr in principle, the substantial reason for my refusal to obey the order to support the TORNADO mission of the German Air Force in Afghanistan was much more far-reaching and complex. In detail, I stated my reasons in my 'declaration plus request' from 15 March 2007 as follows:

> Serving as staff officer in the G3 division, department operations/exercises of the Military Sub District Command IV – Southern Germany – I am responsible for implementing orders for the logistical support of various international missions of the Bundeswehr, inter alia within the framework of EUFOR, KFOR, UNIFIL, ISAF. Since in my opinion these specified missions are based on an adequate legal basis in terms of constitutional as well as international law, I was able to fulfil my tasks with clear conscience up to the present.
>
> However, the latter ... does not apply to the orders concerning the deployment of TORNADO weapon systems in Mazar-i-Sharif.
>
> I hereby declare that it goes against my conscience to support the operation of TORNADO weapon systems in Afghanistan in any way, because I cannot exclude the possibility that by virtue of my own action I will contribute to a mission of the German Armed Forces against which there are grave reservations with respect to constitutional, international, criminal and international criminal law. Within that context, the following deliberations must be stressed in particular:[41]
>
> - The operation of TORNADO weapon systems of the Bundeswehr in Afghanistan inevitably implies that Germany is taking part in military actions that are both illegal in terms of international law and not covered by the NATO treaty, for the reason that
> - the results of reconnaissance sorties accomplished by Bundeswehr TORNADOs are submitted to the high command of the U.S. forces; at the same time despite the restrictions in the ISAF operations plan indicated in

40 Jürgen Rose, *Dienstliche Erklärung vom 1. Mai 2006,* documented in Jürgen Rose, 'Soldaten-Verweigerung gegen völkerrechts- oder grundgesetzwidrige Befehle', *FriedensForum – Zeitschrift der Friedensbewegung*, 3 (2006), p. 9.

41 For details see Dietrich Murswiek, *Organstreit Gauweiler/Wimmer – Antragsschrift vom 9. März* 2007, especially par. C, pp. 16-45 http://www.jura.uni-freiburg.de/institute/ioeffr3/forschung/papers/murswiek/organstreit.pdf.

the resolution submitted, it is not ensured that the results of reconnaissance sorties will not be utilized for other purposes under Operation Enduring Freedom (OEF);

- the war of the USA under OEF is illegal in terms of international law in several respects, that is:
  - it both cannot be justified as self-defence and is not based on a UNSC mandate;
  - concerning its line of approach, especially with regard to the consequences for the civilian population, it exceeds even the authorization granted by the Karzai government;
  - it is incompatible with the rules for the protection of the civilian population stipulated by international law with regard to the so-called collateral damage which it accepts;
  - it violates fundamental principles of humanitarian law concerning the treatment of POWs.
- By resolving to deploy TORNADO weapon systems in Afghanistan, the Federal Government is contributing to warlike operations by virtue of its own actions which are accomplished while resting on a military strategy that is incompatible with the basic principles of the United Nations Charter as well as article 1 of the NATO treaty, and furthermore it entangles the German Armed Forces herein.

Concurrently, I hereby request to be explicitly released from all tasks that could put me at risk to become guilty in the sense exposed above: since my conscience instructs me that it cannot be morally justified under any circumstances at all to break the law – in the present case that is international as well as constitutional law – by virtue of my own action. Such action that is to say would be diametrically opposed to the 'Categorical Imperative' postulated by the philosopher Immanuel Kant, which reads: 'Act only on that maxim whereby thou canst at the same time will that it should become a universal law',[42] or phrased similarly: 'Act so that the maxim of thy will can always at the same time hold good as a principle of universal legislation.'[43] Accordingly, each action which is in compliance with the principles defined by that 'Categorical Imperative' is a priori morally right, while any action that goes against those principles is a priori morally wrong. It is evident and requires no further explanation that one cannot want that breach of law becomes a universal rule under the 'basic law of pure practical reason'[44] – because such volition would be self-contradictory and hence annihilate itself. Not to take part in breach of law by virtue of one's own action hence constitutes an inevitable obligation of moral law, and that is exactly why I shall, as a matter of principle, not take part in actions or support those by virtue of my own action

42 Immanuel Kant, *Grundlegung zur Metaphysik der Sitten* (Hamburg: 1965), p. 42.

43 Immanuel Kant, *Kritik der praktischen Vernunft* (Hamburg: 1974), p. 36.

44 Ibid.

> that imply breach of (international and constitutional) law. In particular this applies to all measures linked with the decision to deploy TORNADO weapon systems in Afghanistan, because these obviously infringe the charter of the United Nations as well as the German Basic Law.
>
> In addition, if I support the operation of TORNADO weapon systems in Afghanistan by virtue of my own action that give rise to grave reservations with respect to constitutional, international, criminal and international criminal law, I shall run the risk of breaking my oath of duty once sworn which urges me inter alia "bravely to defend the law … of the German people", but definitely not to break and trample it. Beyond all question, international law as well as Basic Law are among the central legal norms of the German people.[45]

The refusal to contribute to the logistic support of the TORNADO mission in Afghanistan became the subject of TV coverage, was broadly reported by many radio stations, and could be read in nearly all German newspapers. Not least because of the unforeseen publicity and a constitutional complaint filed with the Federal Constitutional Court,[46] the relevant military authorities promptly decided to move the objector to another division of his department, just as it was settled in the Federal Administration Court's decision regarding the soldier's freedom of conscience. Apart from this, higher commands, including the Ministry of Defence, took no further action on this issue.

Sergeant Major Christiane Ernst-Zettl

The case of the medical orderly Christiane Ernst-Zettl,[47] unlike the cases depicted above, does not refer to the so-called *ius ad bellum*, but to the *ius in bello* or Humanitarian Law.[48] Specialists of international law subsume the latter under the so-called *ius cogens*, that is, law that has to be followed at all times and anywhere.

45 Jürgen Rose, 'Dienstliche Erklärung / Dienstlicher Antrag an Herrn Oberst i.G. Bernhard Frank, Chef des Stabes WBK IV – Süddeutschland –, vom 15. März 2007' (official document, only available on the internet; www.ag-friedensforschung.de/regionen/Afghanistan/tornado-rose.pdf; extracts published in 'Jürgen Rose, 'Aufklären, damit die anderen bomben können. Dokumentation. Antrag des Oberstleutnants Jürgen Rose, von allen dienstlichen Aufgaben bei einem Tornado-Einsatz in Afghanistan entbunden zu werden', in *Freitag – Die Ost-West-Wochenzeitung*, 23 March 2007, p. 7.

46 Dietrich Murswiek, *Organstreit Gauweiler/Wimmer – Antragsschrift vom 9. März 2007* http://www.jura.uni-freiburg.de/institute/ioeffr3/forschung/papers/murswiek/organstreit.pdf.

47 On this point, see Jürgen Rose, 'Die Sanitäterin im Krieg. Der Fall der Bundeswehrsoldatin Christiane Ernst-Zettl und der Bruch der Genfer Konventionen am Hindukusch', in *ZivilCourage*, 1, February/March (2008), pp. 14-15.

48 That law is primarily codified by the four Geneva Conventions from 1949 and the two Additional Protocols from 1977.

The conflict was caused by assigning guard and security duties, that is, protection of military installations, to members of the German Army Medical Corps during the war in Afghanistan. This not only included the custody and defence of army medical service facilities, such as a combat support hospital, which is in full compliance with international humanitarian law, but also implied the comprehensive military protection of the multi-national armed forces' bases in Afghanistan. For that purpose medical orderlies were even employed as machine-gunners, after having been ordered to take off their Red Cross armbands.

Sergeant Major Christiane Ernst-Zettl had been deployed to the surgical company of the medical detachment of the 7th contingent of the NATO International Security Assistance Force (ISAF). During that appointment the Bundeswehr medical orderly was ordered to perform armed service, contrary to international law, when as part of the military protection of Camp Warehouse, the multi-national ISAF garrison based in Kabul, she was requested to undertake personal checks on Afghan women employed by ISAF as local staff. When she informed her superior that she was a non-combatant under humanitarian international law and therefore could not be employed for security tasks, she received a disciplinary fine of EUR 800 and was 'repatriated', that is, ordered back to Germany as a punishment. The soldier's complaint against this treatment was rejected by the responsible 'Military Service Court South' with absurd arguments. Although the complex position of the case under international law resulting from the four Geneva Agreements of 1949 and the two additional protocols of 1977 cannot be described in detail here, the core concern arising from it may be summarized as follows: the questionable nature of the procedure described is that it is likely to deter soldiers from facing up to the legal or moral implications of their actions.

Conclusions

The cases depicted above demonstrate that, if both military superiors and their subordinates are insistently urged to act in accordance with national and international law, it can no longer be ruled out *eo ipso* that those in uniform, whatever their rank, will occasionally refuse to take part in military activities when these are recognizably in breach of national and/or international law or at least are dubious and controversial in terms of national and/or international law. Such cases of selective conscientious objection, which were mainly based on the argument of illegal participation in a war of aggression prohibited under international law, have occurred not only in the Bundeswehr, but also in the armed forces of Germany's allies.[49] These incidents take different forms, ranging from being

49 On this point, see, Ladiges, *Irakkonflikt.* Two prominent cases within the rows of NATO armed forces, namely that of First Lieutenant Ehren K. Watada, U.S. Army, as well as that of Flight Lieutenant Dr. Malcolm Kendall-Smith, Royal Air Force, are depicted in detail in Rose, *Ernstfall Angriffskrieg*, pp. 187-92.

temporarily absent without official leave (AWOL) through permanent desertion or the rejection of individual orders to a complete rejection of war service on grounds of conscience.

Despite all of the differences between the cases of selective conscientious objection described above with regard to the motives of the protagonists and the subsequent reactions of the various military and judicial figures involved, they have one significant aspect in common. This relates to the specific self-perception of the military in modern industrial societies.[50] There it is no longer the unreserved readiness to go to war which characterizes the attitude of the military leadership, but rather caution and reluctance. And there is another factor:

> Whether we like it or not: The military oath…has gradually lost its timeless and indissoluble effect as part of the general change of values. The oath now is no longer sworn to a general or an ideology, but to the nation and its Basic Law.[51]

It is precisely these cases of selective conscientious objection, that is, disobedience to orders in the context of breaches of both *ius ad bellum* and *ius in bello*, that confirm this thesis. Therefore soldiers can no longer be regarded as mere artisans of war 'with high-flying ideals and a blindfold in the national colours before their eyes', as the prominent German pacifist journalist Kurt Tucholsky once wrote, but must be seen as constitutional patriots. In the Bundeswehr this latter type of soldier corresponds precisely with the model advocated above all by the General Baudissin after the Second World War, the 'citizen in uniform' who does not abandon ethical convictions and political ideas even while on military service. But this latter point is not limited exclusively to the military in Germany, because elsewhere too '… the individual soldier, a company, a regiment or even a larger formation have quite often taken to themselves the freedom to decide which orders should be obeyed, and which not.'[52] In addition, as argued above, soldiers are even obliged to refuse to obey orders which are criminal and contrary to international law.[53]

Today the crucial problem is the attitude and the motivation of troops and their identification with the mission.

> If they trust their leaders and accept the 'war aims', then they will be ready for almost anything. But if their attitude differs radically from the view of their (military or political) leaders, this may lead to protests or at least to passivity.[54]

---

50 See Gustav Däniker, *Wende Golfkrieg. Vom Wesen und Gebrauch künftiger Streitkräfte* (Frankfurt am Main: Umschau, 1992) and Jürgen Rose, 'Vom Wesen und Gebrauch zukünftiger Streitkräfte', in *Truppenpraxis*, 4 (1993), pp. 346-51.

51 Däniker, *Wende Golfkrieg*, p. 203.

52 Ibid.

53 See Michael Pesendorfer, 'Handeln auf Befehl – kriegsvölkerrechtlich ein Rechtfertigungsgrund?', Österreichische *Militärische Zeitschrift*, 35/5 (1997), pp. 539-44.

54 Däniker, *Wende Golfkrieg*, p. 204.

The 'refuseniks against wars of aggression', whatever their origin, illustrate strikingly that orders for dubious purposes are no longer being obeyed without question: 'When the legitimacy of the order is not entirely clear and the just cause is not obvious to everyone, the formerly feared instrument ... becomes a recalcitrant rabble'.[55] This means that to an ever-increasing extent, troops can no longer be used for whatever political purpose is desired, and that they can no longer be assumed to willingly participate in military interventions. Admittedly, at the moment the political and military decision-makers have no real reason to fear 'mass disobedience' or even a 'rampant over-use' of conscience. In reality the number of those who refuse orders for reasons of conscience is tiny. Moreover, not in a single instance was the operational readiness or functioning of the armed forces impaired or even touched.

Nevertheless, in the eyes of the authorities the existence of a new type of soldier who is not prepared to obey like a robot, against his or her own conscience and idea of what is right, and is not intimidated by the serious punishments possible under military jurisdiction, must present a serious threat. But as the cases of selective conscientious objection demonstrate, a good deal of pressure is applied to those who refuse to obey orders on grounds of conscience. This starts with the criticism of forbidden political activity, which allegedly calls into question the primacy of the political decision-makers, and culminates in the accusation that it amounts to a challenge to democracy itself. It also allegedly undermines military morale and discipline. The prosecution regularly insinuates that the conscientious objectors only used their conscience as a pretext to hide political motives. And just as regularly, the judicial authorities involved, whether military or civil, refuse to take any account of the reasons put forward to justify disobedience – that is, illegal and immoral wars of aggression and serious breaches of the laws of war, as codified in humanitarian international law. However, the political and military leaders cannot avoid recognizing 'that not only the individual soldier, but even the toughest troop has a soul of his/her own, and a conscience which tells him/her what you can do and what you cannot'.[56] Remarkably, it was a U.S. Army officer, First Lieutenant Ehren K. Watada,[57] who demonstrated clearly that the modern soldier is sometimes very much aware of this, when he declared: 'To stop an illegal and unjust war, the soldiers can choose to stop fighting it ... If soldiers realised this war is contrary to what the Constitution extols – if they stood up and threw their weapons down – no President could ever initiate a war of choice again.'[58]

---

55 Ibid.

56 Däniker, *Wende Golfkrieg*, p. 205.

57 On this point, see in detail Rose, *Ernstfall Angriffskrieg*, pp. 187-9, and Ladiges, *Irakkonflikt*.

58 First Lieutenant Ehren K. Watada quoted from Jeremy Brecher and Brendan Smith, 'Will the Watada Mistrial Spark an End to the War?', *The Nation*, 10 February (2007), http://www.commondreams.org/views07/0209-23.htm

PART III
Conclusions

Chapter 12
Selective Conscientious Objection: Some Guidelines for Implementation

J. Carl Ficarrotta

Of late it has been argued in many places[1] that, in addition to states allowing their citizens to apply for general conscientious objector status, members of any military organization ought to be granted the opportunity to apply, on moral grounds, for excusal from a *particular* conflict, what has come to be called selective conscientious objection. The arguments for this view (some mounted in this collection) are persuasive, and while acknowledging there are many who remain unconvinced,[2] we might plausibly assume here that the case for selective conscientious objection has been made well enough. Rather than rehearsing the current state of the debate, the aim of this essay is to explore how a military might *implement* such a policy. Indeed, deciding that allowing for selective conscientious objection is the right thing to do is only a beginning. Implementing the policy will present a number of vexing problems, problems that will need to be addressed through moral, practical and political lenses.

What follows are suggestions for some high-level guidelines for workable selective conscientious objection procedures. Those guidelines will be developed against the backdrop of two large assumptions: (1) that the military in question is part of a larger political structure that *already* abjures using conscription to fill the ranks, or short of that, at least allows for deferral via general conscientious objection, and (2) that the military belongs to a state that expects its soldiers to obey the laws of war and will punish those who are discovered to have committed war crimes. These may seem at first like overly strong assumptions to make before undertaking this

1 For just a few examples, see: George Clifford, 'Legalizing Selective Conscientious Objection', *Public Reason*, 3/1 (2011), pp. 22-38; J. Carl Ficarrotta, *Kantian Thinking about Military Ethics* (Farnham: Ashgate, 2010), pp. 73-86; Jeff McMahan, *Killing in War* (Oxford: Oxford University Press, 2009), pp. 95-103; Robert L. Muhlnickel, 'Just War Theory and Selective Conscientious Objection', *Review Journal of Political Philosophy*, 6/2 (2008), pp. 1-16; Paul Robinson, 'Integrity and Selective Conscientious Objection', *The Journal of Military Ethics*, 8/1 (2009), pp. 34-46; Jessica Wolfendale, 'Professional Integrity and Disobedience in the Military', *The Journal of Military Ethics*, 8/2 (2009), pp. 127-40.

2 For one formidable argument in this more conservative direction (along with some sharp criticism of academic writers in favour of reform), see, George Lucas, 'Advice and Dissent: "The Uniform Perspective"', *The Journal of Military Ethics*, 8/2 (2009), pp. 141-61.

job, but they are not. This is because in the vast majority of cases where selective conscientious objection is to be implemented, the society will already have those assumed policies in place. Indeed, a moral-political environment with no draft, or at least some provision for general conscientious objection, and a respect for the rules of war is the most likely soil (perhaps the only soil) out of which an enlightened move toward selective conscientious objection might grow.

Of course, we might imagine a country, once upon a time or place, which conscripted its soldiers and expected obedience to even immoral orders, yet wanted to grant its fighters rights to conscientious refusal. But this country would be a strange one, a country experiencing moral development in rather an odd order. To the extent that what follows depends on the staked initial assumptions, it is weakened, but the risk of proceeding is greatly mitigated by the presumed rarity of the alternative cases. Sticking with the initial assumptions about the moral/legal character of the countries in question will also be helpful in narrowing our attention to selective conscientious objection *ad bellum*, rather than refusal *in bello*. Inside the assumptions, refusal *in bello* to illegal orders is already not simply permitted, but expected. Moreover, in the nations we are modelling, there is substantial overlap between the legal and the moral in combat; and when there is not, the most sensible approach is not to introduce yet another category of refusing, but to correct the laws of war and the rules of engagement, reforming inside the already extant legal framework. What is completely missing in our hypothetical country is a legal mechanism for selective *ad bellum* refusal. Implementing that is what will be explored here.

Morally Speaking, How Easy Should It Be to Do?

At first blush, we might think that any procedures we implement for applying for selective conscientious objection should be easy to use. Moreover, there should not be any bad consequences for an application, whether it is successful or not. Why might we think so? Because we also believe there should be no serious obstacles to acting in accordance with one's conscience in such serious matters. Granted, in the vast majority of cases where conscience clashes with the dictates of the body politic, we should rightly be prepared to obey the state; this is an essential part of what it is to be governed and a necessary condition for a functioning state.[3] But many have rightly decided after reflection on general and selective conscientious objection that government ought not to compel its citizens to *kill or be killed* for causes they deem immoral.[4] The lethality in this context presents a special case,

3 For a similar point, see Roger Wertheimer, 'Reconnoitering Combat Moral Equality', *The Journal of Military Ethics*, 6/1 (2007), p. 68.

4 In 1931, Chief Justice of the U.S. Supreme Court Hughes led a luminary judicial minority including Holmes, Brandeis and Stone in making this same point: 'Much has been said of the paramount duty to the state, a duty to be recognized, it is urged, even though it

an exception to the important general principle of submitting to political authority. Avoiding a difficult-to-use selective conscientious objection procedure, or one that resulted in de facto punishments for using it, would be strongly advisable for the same reasons we want a procedure in the first place: we want to give soldiers an out when they cannot participate in good conscience, not discourage them from so deciding or punishing them if they do.

Besides, taking the decision to apply for selective conscientious objection would undoubtedly be extremely difficult, difficult enough without the procedure itself adding to the problem. We know the powerful forces at work on soldiers to believe their causes are just and to fight even when they might have serious reservations. Patriotism, group solidarity, self-justification mechanisms of various sorts, fear of being branded a coward, strong desire to continue serving in other capacities (and to continue one's career), thoughts that one's own refusal would not make any difference in whether the war is fought, and many other influences make it remarkable that *any* soldier comes to see his nation's cause in a particular conflict as unjust and immoral to pursue. If, in the face of all these countervailing pressures, a soldier comes to the decision that he or she ought not to fight, then we should not make acting on the decision any more vexing than it has to be.

How easy could we make such an application? Let us begin by sketching the easiest (for the soldier) procedure we can imagine. Make it a simple form. While commanders in the applicant's unit should be made aware of the application, and should be solicited for their opinions about it, exclude immediate higher-ups from the decisions that follow, so as to avoid undue command influence on the process. Have a military specialty assigned responsibility for assisting soldiers in the process, such as military defence lawyers, or chaplains, or some other group that we can assume will be able to advise and assist with the member's moral and legal interests in mind. Establish reasonable, attainable standards of proof that the applicant genuinely wants a deferral on deeply held moral or religious grounds. Have the procedure for approval streamlined and made at a reasonably low level of command (or with a designated civilian authority), so as to reduce

---

conflicts with convictions of duty to God. Undoubtedly that duty to the state exists within the domain of power, for government may enforce obedience to laws regardless of scruples. When one's belief collides with the power of the state, the latter is supreme within its sphere and submission or punishment follows. But, in the forum of conscience, duty to a moral power higher than the state has always been maintained. The reservation of that supreme obligation, as a matter of principle, would unquestionably be made by many of our conscientious and law-abiding citizens. ... There is abundant room for enforcing the requisite authority of law as it is enacted and requires obedience, and for maintaining the conception of the supremacy of law as essential to orderly government, without demanding that ... [moral and religious conviction be] subordinate to allegiance to civil power. The attempt to exact such a promise, and thus to bind one's conscience ... has been the cause of many deplorable conflicts. The Congress has sought to avoid such conflicts in this country by respecting our happy tradition. In no sphere of legislation has the intention to prevent such clashes been more conspicuous than in relation to the bearing of arms' (*United States v. Macintosh*, 283 U.S. 605).

the time needed for a decision. Reassign the applicant, consistent with the needs of the service, to other duties that are not punitive in character (and which might, by the way, include combat duty in a different conflict). Do not take away any privileges, benefits or awards the soldier would otherwise be entitled to, except those he would have earned specifically for serving in the objected-to conflict. Last, mask the application for selective conscientious objection status from future promotion and assignment boards.

This proposal, as sketched, gives us a general framework for a system that sets up almost no barriers to an objector other than the already nearly insuperable social and psychological ones. From the moral point of view, it assumes the best about potential applicants and makes it as easy as possible for them to do what is, by their own lights, the right thing. But it remains to be seen whether this is, all things considered, what we should want.

Practically Speaking, How Easy Should It Be to Do?[5]

Of course, there are certainly many who think implementing *any* mechanism for selective conscientious objection, much less an easy one, will result in multiple, intractable and catastrophic problems. Careful examination of these potential problems with selective conscientious objection reveals that they would not be serious enough to recommend against moving forward with an easy-to-use application procedure.

To begin, might it be that an easy-to-use application for selective conscientious objection would encourage cowardice or give cowards a previously unavailable way out of doing their duty? If this is true, then it would be a serious worry. Abandoning one's fellows as they go into combat to face injury and death might end up being, if done for the wrong reasons, a moral failing of the very first rank. Desertion because of fear[6] is an occasion for shame even if it is done during the heat and terror of combat; to opt out because of fear before combat even begins would be to display an even more egregious kind of cowardice. The spectre of past deserters ought to trouble us greatly and we definitely do not want to make it easy to do something so seriously wrong.

We might appreciate the worry, and grant it is not an unreasonable one to raise, and nevertheless not think cowardice will cause important numbers of people to seek (disingenuously) selective conscientious objection status. Here is why. Where there is conscription or mandatory registration for potential conscription, we may assume cowards are already feigning a *global* objection to war, and have already sought general conscientious objector status. Where there are volunteer militaries, we will find almost no full-bodied cowards anyway, since they will

5 In the largest part of this section, I will be repeating many objections and replies to selective conscientious objection that originally appeared in Ficarrotta, *Kantian Thinking*.

6 And we should notice, any lesser motivation, like convenience.

not have joined; becoming a soldier is definitely not the career of choice for those hoping to avoid danger.[7] Take the current US military as a strong counterexample to the worry about cowards in volunteer militaries: in that volunteer military, in a difficult, dangerous and morally controversial war in Iraq, during a time when deserters typically were merely discharged from the military (rather than shot or imprisoned), the number of military professionals who fled the military as deserters was miniscule.[8] And this small number almost certainly included not only cowards but also those who left on principle, making the 'coward count' even smaller. Indeed, it is one of the greatest fears of any soldier that he will not master his own fear, will fail in the face of danger, and have to live with the shame of cowardice. We might predict that deserter numbers will be small.

So we will most certainly not find many volunteer and professional soldiers who will deliberately endure the shame and contempt of their fellows that would result from a careful decision to give in to fear, one made ahead of the danger. In this population of military professionals, only grave moral concern (even in the face of the concomitant risk of being *thought of* as a coward) could plausibly account for such a considered decision. Real cowardice and desertion are far more likely in armies wherein soldiers are forced to fight or are not properly trained and socialized into units. In all, on this point we should be confident that selective conscientious objection would function not as a revolving door for cowards, but rather as a rarely used escape hatch for thoughtful and morally sensitive military professionals. And leaving or avoiding a fight because one thinks it is immoral is not the same thing at all as leaving or avoiding a fight simply out of fear.

This all being said, one might have a lingering concern about cowardice that ought to be taken up. We all frequently engage in rationalizations, in every aspect of our choices and behaviour, from small matters to serious ones. The true springs of our motivations are often enough hidden from our own view.[9] This will be

---

7 Also, it is probably safe to assume that even in a bad economy, financial incentives will not be sufficient to induce cowards to sign up for possible combat.

8 Throughout the long years of the wars in both Iraq and Afghanistan, the number of deserters has fluctuated somewhat depending on a number of factors, but has in every year been significantly less than one per cent of the force. For just some of these reports, see: Sig Christenson, 'Army Desertions Drop as Economy Fizzles', *San Antonio Express-News*, 24 September 2011, available at http://www.mysanantonio.com/news/military/article/Army-desertions-drop-as-economy-fizzles-2186627.php accessed 30 January 2013; Bill Nichols, '8000 Desert During Iraq War', *USA Today*, 7 March 2006, available at http://usatoday30.usatoday.com/news/washington/2006-03-07-deserters_x.htm accessed 30 January 2013; and Rod Powers, 'US Military Desertion Rates', US Military section in *About.com*, 13 March 2005, available at http://usmilitary.about.com/od/justicelawlegislation/a/desertionrates.htm accessed 30 January 2013.

9 For an accessible summary of a great deal of social scientific and psychological research on this point, see Carol Tarvis and Elliot Aronson, *Mistakes Were Made (But Not by Me): Why We Justify Foolish Beliefs, Bad Decisions, and Hurtful Acts* (San Diego: Harcourt, 2007).

especially prevalent in cases where we choose in ways that might be, by our own lights, immoral. Rather than hold ourselves up to moral criticism, we will much more likely tell ourselves and others a different story as to what we are up to. We will rationalize, re-describe, self-justify, anything it takes, to avoid having to think of ourselves as moral failures. In the case of cowardice, couple self-loathing and the stigma others will brand us with, and we have the perfect conditions for succumbing to self-delusion. How can we be sure that the selective conscientious objection applicant is not fooling *himself* about whether he is a coward rather than a moral exemplar? To soften this interesting concern, we can begin by resurrecting a point already made: someone inclined to do this to avoid combat *will have done it already*. Simply avoiding military service altogether, whether by not joining or via general conscientious objection, was available all along. If we are concerned that a looming deployment might make the risks (and the attendant fears) more vivid, fair enough. Perhaps application for selective conscientious objection should require some time spent with an advisor, counsellor or member of the clergy, for the purpose of probing this possibility. Some couples take pre-nuptial counselling before going through with a wedding, to make sure they are marrying for the right reasons. A similar reflection with a third party would not unduly burden an applicant for selective conscientious objection. Put the requirement into a check list on the form. But to prohibit selective conscientious objection altogether is too blunt an instrument. Similarly ham-handed approaches to preventing self-deception would speak against allowing general conscientious objection too, along with choosing one's own job, voting for one's favourite political candidate, marrying whom one pleases, or any other important life choice where self-deception can queer the outcome.

Here is another objection to selective conscientious objection that might affect how it should be implemented. Perhaps if the military or its membership were allowed to decide they will *not* fight, independent of the pronouncements of their political leadership, this might make them more inclined *to* fight even when their political leadership says they should not. If military people are allowed to decide when not to fight, they might be tempted to think they know better about when to fight, or even in the extreme whether they should take over the government. If this worry has merit, it seems to encourage, or at least make too much room for the likes of Oliver North, Douglas MacArthur, societies unduly influenced by the military branch, or even military dictatorships. But selective conscientious objection is a mechanism *only* for a legal, openly exercised option for the individual soldier, not for the military as an institution. And concerns about political control over the military have always and properly been about whether the military would exert unchecked power as a larger institution, not whether individuals would on grounds of conscience *fail* to participate in wars or insurrections. Selective conscientious objection is clearly intended to present a legal option *not* to fight, rather than a permission to take more liberties in the opposite direction. There is no good reason to believe the one (a way to opt out) will lead to the other (a tendency to take over).

Again, we should not confuse refusing to fight in a conflict overall with refusing to perform a particular mission or action on moral or legal grounds. The latter is a particular refusal *in bello*, in order to comply with the laws of war or avoid participating in an action that would shock the conscience; the other is an *ad bellum* judgment made at the level of the participant. Yet consider a parallel between the two. Moral militaries *already* expect their soldiers to refuse to participate in atrocities and punish those who fail to refuse. Of course, you had better be right about the problem, or you will face discipline of a different kind. But notice that the permission, indeed, the demand, to refuse in the combat (without so much as an application form!) has not led to a breakdown of command authority.[10] In the selective conscientious objection case, we change only the level of the objection, from *in bello* to *ad bellum*. We should not expect any huge difference in the effect such objectors will have on legitimate authority, of either commanders or states. This is no reason to make the application more difficult.

Another objection, a traditional one voiced from Vitoria onward,[11] claims that if soldiers were allowed to pick and choose the wars in which they fought, it would weaken their military and put their nation's security at grave risk. This worry about insufficient numbers of reliably available troops runs exactly counter to all the social forces already mentioned. Only the *patent* injustice of a conflict could ever hope to overcome the strong tendencies there are in the ordinary person to go along with the decisions of his country. Kant thought that only a republican form of government would check rampant war-making between states, since the sovereign would be answerable to the people via the republican structures of state; massive casualties and costs for ill-advised wars would not be sustainable.[12] Perhaps we should take the same thought a bit farther. If selective conscientious objection is available, the sovereign would also be required to convincingly justify directly to its citizenry and soldiers the morality of the wars it aims to make. If we assent to the modest claim that fewer unjust wars is a good thing, we have all the more reason to endorse an easy-to-use selective conscientious objection procedure. Sadly, we should also suspect that even in patently unjust wars there would still be plenty of people available to fight. Should the war ever be so unjust or ill-advised that we see mass refusals, we should welcome any government's frustration.

In countering this worry about military strength, we also find an unlikely bedfellow in Walzer,[13] who in his fine discussion against conscription introduces

10 Notice that some SCO advocates have concluded that refusal to participate is not just a right, but a duty, just as it is in the combat. See McMahan, *Killing in War*, and Wolfendale, 'Professional Integrity'.

11 Francisco de Vitoria, 'On the Law of War', in Anthony Pagden and Jeremy Lawrance (eds.), *Vitoria: Political Writings* (Cambridge: Cambridge University Press, 1991), 2.2, §22 and §25.

12 Immanuel Kant, 'Perpetual Peace', in H. Reiss (ed.), *Kant: Political Writings*, 2nd Enlarged Edition (Cambridge: Cambridge University Press, 1991), p. 100.

13 Unlikely, since he is a traditionalist when it comes to assigning *ad bellum* decisions to the state, not the soldier. See Michael Walzer, *Just and Unjust Wars: A Moral Argument with Historical Illustrations* (New York: Basic Books, 1977).

a distinction between mere citizens and officers of the state. When the state commands, it does two things. First, it requires its citizens to obey. But second, it also appoints people to *enforce* the law. Refusal to obey laws or policies is something the state finally cannot tolerate. However, the state need *not* require a particular person to carry out or enforce state decisions: a refusal to become a state servant, official or soldier 'does not prevent the state from carrying out its policies. The state can always find other servants.'[14] The state can easily tolerate or grant refusals of service and not collapse or cease to perform its essential functions, including its military functions. Laws could be enforced and wars fought by citizens who are in favour of, indifferent toward or weakly opposed to the laws and/or wars.[15] 'The state can survive and flourish without a conscript army (as it does without a conscript bureaucracy).'[16] Walzer is right, and likewise, in a particular conflict, there will always be others to take the place of a conscientious objector (unless, and again, the war is so obviously immoral that it clearly ought not to be fought anyway). None of this should lead us to make application for selective conscientious objector status any harder.

Indeed, rather than reducing a nation's military strength, the ready availability of selective conscientious objection might actually have some benefits. First, more people might be inclined to join a military in which they were not required irrevocably to surrender their consciences.[17] Second, a military in any given war comprised of only volunteers who *also* had no moral compunctions about the conflict might well have higher morale and be more effective in that war.

Perhaps selective conscientious objection would be a bad idea because it would almost certainly, in our own day, be administered inequitably. This worry has it that in the case of an unjust war, only the better educated members of society, those already privileged in so many ways, would be likely to realize that the war is unjust and ask for selective conscientious objection status. Hence, mostly people with poor educations, tending to come from the lower socio-economic classes, would be doing the fighting. We could reasonably doubt whether the outcome predicted by this objection would obtain. But if inequity is the result of otherwise free choices, the fix for the problem would be to raise the general level of education (which is so essential to the proper functioning of any decent society), not to require people to fight and kill in violation of their consciences. Surely that coercion would be morally worse than the possibility of inequity while we wait to improve general education. Still, in implementing a selective conscientious objection procedure, we should do what we can to inoculate against this potential problem. We should make sure soldiers know that they have such an option and that they understand what it allows, perhaps by briefing them during basic training; and as we have

---

14 Michael Walzer, *Obligations* (Cambridge, Massachusetts: Harvard University Press, 1970), p. 136.

15 Walzer, *Obligations*, p. 137.

16 Walzer, *Obligations*, p. 145.

17 This is a point I owe to Rick Furtak of Colorado College.

suggested already, we should also make sure there are functionaries assigned the duty of assisting soldiers in the application process.

Contrary to what has been said so far, some might worry that an easy-to-use application for selective conscientious objection would have an unintended effect of making it too *easy* (rather than harder) for adventurous governments to wage immoral wars. If no one is drafted, no one is forced to serve generally, *and* no one is unduly pressured to serve in any particular war, then part of a citizenry and their families lose their strongest motivation for protesting an unjust war: they are no longer in the same boat with those who will mistakenly die or be hurt for a bad cause. And all this might make it easier politically to fight an immoral war. This objection is interesting but misses the mark. Of course, those who have lots at stake personally will protest an injustice more quickly and more vigorously than those who do not. But the history of states, especially functioning democracies, plainly shows there are also sufficient numbers of citizens who will fight evil being done in their name, even when they are not directly harmed or benefitted. And again, even if the number of unaffected people that would protest turns out to be smaller than we hoped, and the free choices to opt out make it a perhaps a *little* easier to wage unjust wars, the coercion to fight and kill contrary to one's conscience would be morally worse.

Last, consider a concern based on the timing of selective conscientious objection applications. We might grant that allowing for selective conscientious objection is the right thing to do, and even concede that an application for it prior to deployment would be workable. But surely, one might think, if a soldier changes his mind about the morality of the conflict after joining the fight, it will be much more difficult to put procedures in place in a combat zone. This, it might be argued, has to be a limit on applying for selective conscientious objection status: before deployment, you have your chance to say no, but once in theatre, it is a luxury we cannot and should not provide. While initially plausible, this approach would be wrong. In terms of justifying the refusal, nothing changes because the fight has started. Soldiers might learn things about the morality of the war that were not apparent before deployment (but not, by the way, discovery of *jus in bello* problems; these are separate issues). And the history of war shows how easy it is for causes to change as a conflict unfolds: there is often what has been called 'mission creep' and the ends these modified missions pursue may not be on all fours morally with the initial *jus ad bellum* story. Moreover, the idea that an application simply could not be made or processed in a combat zone is nearly absurd. Of course, it is more difficult to conduct administrative business while deployed to a combat zone, yet it is routinely conducted nonetheless. Leave forms are processed, supply requests are put in, humanitarian reassignments made, medical excusals granted, performance reports written and dozens of other administrative tasks like them are accomplished very day. Communication systems now common in modern militaries make solving such problems even more tractable. There is room here as well for an easy-to-use selective conscientious objection application.

Politically Speaking, How Easy Should It Be to Do?

So let us take stock. For our purposes here we have assumed the opportunity to apply for selective conscientious objection is morally appropriate. We have also assumed that such an application process is being developed for a modern society that also has no draft (or at least permits general conscientious objection) and demands that its military members follow the laws of war in combat. We saw that deciding to apply for selective conscientious objection status would be, in moral and psychological terms, a very difficult thing for a soldier to do. Because of this, the application process itself should be as easy as possible, so as not to put up any additions to the already formidable roadblocks. We sketched one possible version of such a process, considered a litany of worries about how and why it might not be practically workable, and concluded the worries are not decisive. In the ideal, a simple, easy-to-use and non-punitive application process is the right way to go.

To many, this recommendation will seem counter-intuitive, perhaps even a bit of unmitigated lunacy. There will likely be a very large number of career military officers, politicians and citizens who will be unconvinced by the arguments made here. Their misgivings will be misplaced, and with more reflection on the evidence, and perhaps some experience with a selective conscientious objection application system, they would come to change their minds. But how shall we get from here to there? Without a political will sufficient to make these reforms, *they will not happen.*

One, and maybe the only, way to proceed here is incrementally. We will have to craft whatever initial foray into selective conscientious objection we offer in a way that assuages fears. And that will mean the application process will be more difficult, and approval will be harder to get, than is ideally morally fitting. What we will have is an example of Walzer's doctrine of dirty hands in the political realm, enunciated so well in 1973:

> [A politician] wants to win the election, someone says, but he doesn't want to get his hands dirty. This is meant as a disparagement, even though it also means that the man being criticized is the sort of man who will not lie, cheat, bargain behind the backs of his supporters, shout absurdities at public meetings, or manipulate other men and women. Assuming that this particular election ought to be won, it is clear, I think, that the disparagement is justified. If the candidate didn't want to get his hands dirty, he should have stayed at home; if he can't stand the heat, he should get out of the kitchen, and so on. His decision to run was a commitment (to all of us who think the election important) to try to win, that is, to do within rational limits whatever is necessary to win. But the candidate is a moral man. He has principles and a history of adherence to those principles. That is why we are supporting him. Perhaps when he refuses to dirty his hands, he is simply insisting on being the sort of man he is. And isn't that the sort of man we want?[18]

18 Michael Walzer, 'Political Action: The Problem of Dirty Hands', *Philosophy & Public Affairs*, 2/2 (1973), p. 165.

In our present case, we believe we know what a morally appropriate policy looks like, but moving to it straight away would be politically impossible. And is not moral and political reform most often like that? Big changes, even for moral improvement, naturally seem to most people too strange to be right, too radical to be embraced. What we have already done for a long time carries with it tremendous psychic momentum. And so the reformer must take the long view, especially at the outset. Often, if a reform is to be eventually successful, we must take smaller steps, each one shortened by a moral compromise. While we may not have dirty hands in Walzer's sense, we will at least have muddy feet, doing less than we ought in each move toward the ideal. Martin Luther King famously invoked the claim that justice delayed is justice denied, and it may appear that this recommendation is in conflict with this sage advice. But in King's case, what he was demanding (racial equality under the law) *was* politically possible[19] and therefore delay was intolerable. But in the case of selective conscientious objection, in many (if not all) countries a less than morally perfect procedure will be the only one that has a chance; and in this context, almost any procedure at all is superior to no procedure, and will serve to gradually and incrementally inure critics to the changes.[20]

What will have to give? Maybe a lot at first, and there is plenty we could put on the table to make a procedure more politically palatable. Compared in Table 12.1 (below) is a vision of the perfect with what we may have to live with.

Just how far it will be necessary to go to get the ball rolling will depend on the political context, which will vary from country to country, or from year to year in any given country. Naturally, proponents of selective conscientious objection should try mightily to keep as much easy usability in place as is possible. But even if a first-pass implementation of a selective conscientious objection procedure carried with it all the restrictions on the right hand side of the table, again this would be better than having no procedure at all, and could serve as an important first approximation to a morally better future. Here finally we can throw our lot in with Dr. King: on this more modest proposal for political action (implementing whatever manner of selective conscientious objection we can manage to get), wherever it is needed, there should be no delay.

Note that not all the compromises on the right hand side of our table are morally equivalent. It would be helpful to evaluate them in terms of how palatable it would be to accommodate them. To begin, there is a clear difference between the elements that make it harder to apply and those that make it harder to get approval. Compromises that make it harder to apply (most obviously, making the form unnecessarily complex, providing no assistance and sending it to inappropriately high levels of command for approval) are morally bogus. While opponents of

19 Evidence for this claim relies on another adage: the best evidence for something's possibility is that it *actually happens*. The civil rights legislation he supported actually passed, hook, line and sinker.

20 Take as an example of this gradual process the U.S. policy of Don't Ask Don't Tell with gay and lesbians, followed by a complete lifting of the ban.

Table 12.1 Political give and take

The Morally Perfect (and the enemy of the good?)	**The Morally Imperfect (but on the table to make selective conscientious objection politically feasible)**
Make it a simple form.	Make the form unnecessarily thorough, legally technical, with heavy coordination and lots of supporting documentation. Make it difficult enough to deter anyone who is not deadly serious.
Applicant's immediate commander does not make or have veto power over the decision.	Require the immediate commander's approval.
Military assigns someone responsibility for assisting soldiers in the process.	Do not commit any military resources to assisting the applicant in the process.
Establish reasonable, attainable standards of proof for applicant's motivation.	Establish very high standards of proof, such that they can rarely be met by someone already in the military.
Procedure for approval streamlined and made at a reasonably low level of command.	Approval at high level of command, requiring complex and time-consuming staff support.
Reassign the applicant, consistent with the needs of the service, to other duties that are not punitive in character (and which might, by the way, include combat duty in a different conflict).	Require the successful applicant to leave the military. Alternatively, allow the applicant to stay in, but whether successful or not, put no restrictions on the next assignment (and along with whistleblowers, let the retributive or punitive chips fall where they may).
Do not take away any privileges, benefits or awards the soldier would otherwise be entitled to, except those he would have earned specifically for serving in the objected-to conflict.	Strip the successful applicant of all military privileges and benefits. No fighting, even in one conflict, means no benefits.
Mask the application for SCO status from future promotion and assignment boards	Even for the unsuccessful applicant, promotion boards and assignment detailers will know about the application.

selective conscientious objection might view such requirements favourably, since they might have the effect of reducing applications, they are transparently and objectionably obstructionist: they skirt the more substantive issues, lay down a gauntlet of purely bureaucratic wickets, and would undermine everyone's faith in the fairness and forthrightness of the command structure. Compromises that remove protections against retributions for even applying are no better. By making selective conscientious objection available, the government says at the same time that application is a legitimate choice for a service member to make, perhaps even a right. Failing to protect applicants against backlash has the institution speaking, as it were, out of two sides of its mouth, again undermining moral confidence in the military. Accommodating these proposals should be at the bottom of the list of things we are willing to do.

Measures that make it, on principle, harder to get *approval* might conceivably be above board and even morally plausible. Proponents and opponents of the policy might meet in the middle, hashing out approval and disapproval criteria, and deciding within reason who shall apply those criteria. Take, for instance, a requirement for the immediate commander's approval. Commanders almost certainly will not want soldiers leaving their units, and almost certainly will not agree with the applicant as to whether the war is sufficiently unjust to warrant refusal to participate. Still, it is nonetheless possible that commanders will be able to recognize reasonable disagreement on a tough issue, and with their direct knowledge of the soldier, make a fair judgment before sending an application forward. Commanders might also be reasonably expected to refrain from treating applicants shabbily afterward, especially if there are some regulatory safeguards in place. The partisan for the easy-to-apply approach will not be completely happy, but may be willing to live with this.

Who should have the final approval for an application and by what criteria should it be judged? Perhaps staff officers inside the personnel system, ones trained in the selective conscientious objection procedure, could provide the last word; this would certainly be one of the easiest ways to proceed, for both the applicant and the service. Compromises in the direction of controlling this more tightly could require more senior officers in the chain to approve. Would a panel or board be more appropriate?[21] Probably not, since they are more costly and cumbersome, and the thorough review and deliberation that a board provides would not likely be required for this kind of decision (except perhaps at the level of an appeal). Someone might object that this in not in line with how many countries go about granting general conscientious objection (where panels often *are* used), but this need not detain us for at least two reasons. First, there may be things that must be sorted out in a general conscientious objection hearing that will not concern us with a selective conscientious objection application, like evaluating evidence for the sincerity of claims about deep-seated and long-held moral convictions. Second, maybe the typical general conscientious objection application should be *easier*. In any event, we are looking to design a procedure for selective conscientious objection from the ground up, one that makes sense and is politically feasible; there are enough things to worry about in that project without accommodating an unwarranted desire to be in lockstep with general conscientious objection procedures.

The foregoing remarks imply that the approval/disapproval decision will not be that difficult. Should we be confident in that? Indeed, just what sort of approval criteria are we imagining? When would an applicant be turned down? The most

21 See for instance, Robinson, 'Integrity', pp. 44 and 46.

liberal approach would require very little from the applicant inside the procedure: fill out the form and attest to the fact that he does not wish to participate in the war for moral reasons. But to make selective conscientious objection more appealing to opponents, we might require more. For instance, the applicant might be required to make clear that she is not refusing on merely political grounds. On one hand, the moral and the political overlap extensively and unavoidably, and one might be tempted to think that all political reasons will bottom out eventually in moral ones. But on the other, it would not be unreasonable to turn down applicants who believe that a war is, say, imprudent because it is too expensive, or not in the (non-moral) national interest. It would also be reasonable to restrict excusals to direct participation. Imagine someone wants to be excused from serving not just in theatre, but from all military service, since all service provides at least indirect support to the fight. The end of this road could find us refusing to pay taxes on grounds of selective conscientious objection. A more restrictive approval process, one that should be considered in any compromise, might insist on excusing only direct and not indirect contributions to the objected-to war.

If there is evidence that the application is actually being made out of fear, this could also be a reason for disapproval. But as has already been discussed, we will not have to be especially worried about cowardice. Inside our initial regulatory assumptions, the selective conscientious objection applicant already volunteered for military service, and could have previously applied for general conscientious objection or simply not enlisted. Moreover, in many militaries the members have important influence over the specialty they will enter, and those that did not want to fight out of fear would not have joined or remained in the combat specialties. But serial applications to avoid combat might provide evidence for an actual motive of fear (or worse, evidence for the essential injustice of his state). Reasonable compromise could allow for disapproval if there is compelling evidence of cowardice, however uncommon it may be in this context.

Sagi and Shapira make a helpful suggestion for making out the difference between civil disobedience, which is done publically and aimed at changing a law or policy, and conscientious objection, which will be undertaken more or less privately and aims only at excusing the applicant.[22] They are quite right that it would be reasonable for a state to require its selective conscientious objection applicants to restrict their purposes to the narrower moral aim of refusing service and to leave broader political agendas out of the process. This concern gives us good reasons to be especially careful and discreet about applications from senior officers; if it was it to be made public that generals were applying for selective conscientious objection, it would almost unavoidably have a political effect that would at least resemble civil disobedience.

Unhappily, after the helpful caution about civil disobedience, Sagi and Shapira go on to introduce a tortured distinction between objections based on (universal)

22 Avi Sagi and Ron Shapira, 'Civil Disobedience and Conscientious Objection', *Israel Law Review*, 36/3 (2002), pp. 184 and 217.

moral principles and merely privately held values.[23] They think their so-called moral conscientious objector is not to be countenanced, since his appeal to universal principles allegedly undermines the state's moral standing. But they think an appeal to a clash with merely private values allegedly does not undermine the state and we may make room for refusal, based on respect for autonomy, in those cases when the objector can 'no longer cope'.[24] This distinction is not sustainable. Moral judgments by their nature appeal to principles the speaker takes to be universal. And appealing to merely privately held values does not carry near the justificatory force needed and is shockingly permissive: after all, cowardice or selfishness might be privately held values. No, in selective conscientious objection, as in general conscientious objection, what we are properly doing is respecting the autonomy of the objector on a tricky *moral* matter *about which we disagree* (and, we should add, only inside the bounds of reasonable disagreement and only inside the narrow context of killing and being killed). Trying to sort out an additional universal/private vector would be wrong-headed and unnecessary.

Conclusion

If, as we assume for purposes of this essay, permitting selective conscientious objection is the right thing to do, then it should be an easy thing for which to apply. In a modern military, powerful social and psychological forces will make it hard enough for a fighter to abstain from participating in war that is, by his or her own lights, immoral. In matters of life and death, further procedural bars are not only unnecessary, but unconscionable. The worries about the moral and practical consequences of implementing selective conscientious objection as a policy are likely rooted in an antiquated understanding of military life, one in which uneducated, mistreated soldiers are bluntly pressed into service via conscription, and the officers need pistols not to engage the enemy, but to shoot deserters. The character of the military we are stipulating for now (and which we hope approximates to what there is in our own day) is a much different and much better one. Still, it will come as no surprise if introducing a workable procedure will require practical and political compromises. So be it. A partial or imperfect selective conscientious objection procedure is preferable to no procedure, especially if it serves as a way station to something better. But along the way, reformers and their opponents alike ought to be mindful of the moral high ground in their disagreements, and to the extent possible, compromises made should not amount to acceding to simple bureaucratic obstructionism.

23 Sagi and Shapira, 'Civil Disobedience', pp. 184-5.

24 Sagi and Shapira, 'Civil Disobedience', p. 185.

Chapter 13
War Resisters in the US and Britain – Supporting the Case for a Right to Selective Conscientious Objection?

Andrea Ellner

Introduction

This chapter examines how far evidence provided by current members of the US and British armed forces or veterans of the wars in Iraq and Afghanistan who became war resisters refutes arguments against the introduction of a right to selective conscientious objection for members of All Volunteer Forces. Although many, including some resisters, assume that having volunteered means one cannot conscientiously object to war, a right to universal conscientious objection exists in the US and Britain. Yet, this is not an option for those who developed conscientious objections to the prosecution (*jus in bello*) and their government's strategic rationale and justification (*jus ad bellum*) of a specific war. The chapter starts with a brief examination of why this is the case and shows that many resisters chose illegitimate ways to separate from the military.

Without suggesting that conscientious objection is the prime motivator for going Absent Without Leave (AWOL) or deserting, it argues that steep rises in such incidents after 2003, particularly in Britain, combined with the revival of resistance groups, especially in the US, indicate that there have been problems to which a right to selective conscientious objection may be a partial solution. The evidence examined after a brief survey of arguments against a right to selective conscientious objection stems from resisters of whom many, though not all, had gone AWOL or deserted at some point. They have publicly explained their ethical objections to the wars, especially in Iraq, and are thus self-selecting as would-be candidates for selective conscientious objection status.

The analysis examines this evidence through engagement with arguments forwarded by other contributors to this volume against introducing a right to selective conscientious objection; further linkages to their contributions are also highlighted in the conclusions to this volume. This chapter focuses particularly on Skerker's conception of individual moral agency in and corporate responsibility of the military. In conjunction with this, it suggests widening the context for the debate on a right to selective conscientious objection. Skerker, Fisher, and Bergeron argue that, because enlisted personnel have volunteered they have knowingly entered into a contract which obliges all members of the armed forces to execute the orders given by the legitimate authority, that is, the democratically elected government. They reject a right to selective conscientious objection as long as the order to go to war is legal or operations are conducted within the Law of Armed Conflict. This chapter argues that we should question the assumption that volunteering equates to making a free choice to join based on recruits understanding and accepting that in service they may be expected to engage in operations that contravene their ethical principles and against which they have no right to object unless orders are manifestly illegal.

Here, and with regard to the concept of corporate responsibility, the context of counter-insurgency operations is particularly relevant. As will be shown, corporate responsibility does not necessarily facilitate service members absolving themselves from personal guilt resulting from ethical conflicts they experienced in war. Individual service members have different ways of dealing with the psychological consequences. Publicly outspoken resisters have, to some degree, externalized their conflicts and transformed their regrets or guilt into public opposition to the wars or into assistance to veterans or other resisters still in service. There are, however, also other responses. In 2009 Litz et al., health professionals working with the US military, identified the phenomenon of moral injury. All wars, they observed, confront service members with ethically problematic choices, but 'counterinsurgency, guerrilla warfare, especially in urban contexts poses greater risks than other wars that military personnel encounter, as 'these types of wars involve unconventional features [...] that produce greater uncertainty, greater danger for noncombat troops, and generally greater risk of harm among non-combatants'.[1]

The evidence presented throughout will show that many experiences which presented resisters with ethical conflicts or caused them to object and absent themselves were closely linked to this operational context. It will also show that military leadership played an important role in setting parameters, such as the Rules of Engagement, which enhanced or even created such pressures. The final section of this chapter then explores, in place of summative conclusions, to what degree leadership and changes in military culture might help reduce moral or ethical conflicts for individual service personnel and what role a right to selective conscientious objection might play in facilitating such change.

The 'All or Nothing' Problem of Conscientious Objection

Britain and the US acknowledge in law that volunteers may develop conscientious objections to war during military service. In the US eligibility criteria have been broadened since the Vietnam era. Applicants may hold traditional religious

1 B.T Litz et al., 'Moral Injury and Moral Repair in War Veterans: A Preliminary Model and Intervention Strategy', *Clinical Psychology Review*, 29 (2009), p. 696.

belief in a supreme being or an overarching set of ethical or moral beliefs, but cannot object on political, sociological or purely personal ethical grounds. They have to be sincere and object to all war.[2] The Non-Governmental Organization *Conscience and Peace Tax International* has described Britain's record on universal conscientious objection as 'on the surface (…) exemplary'.[3] Deakin's account in this volume of the British approach to selective conscientious objection shows that such cases have traditionally been, and are still, handled pragmatically. This approach minimizes publicity, political embarrassment and the potential for resistance to spread amongst service members. It is indeed difficult to establish an accurate picture of the number of applications and success rates in contemporary Britain and the US.

The official US figure of 425 applications, with 224 approved, between 2002 and 2006, is widely regarded as unreliable and 'extremely low'.[4] The British Ministry of Defence (MoD) keeps no statistics on applications and only released figures of approved applications following requests under the Freedom of Information Act. According to this information, the largest number approved in one year was six.[5] In both countries those advising or providing legal representation for service personnel estimate that the true number of individuals who sought to leave service during the Iraq War on grounds of conscience was significantly higher but several factors obscured the picture.

According to Bill Galvin of the US Center on Conscience and War, universal conscientious objection applications dealt with at a local level are not recorded and many conscientious objectors 'find some other way to get out. They never apply for conscientious objector status because it's not the easiest way. Some of them go AWOL, some can prove medical reasons or some may challenge their enlistment agreements.'[6] British military lawyer Gilbert Blades also stated that

2 Daniel H. Shubin, *The Conscientious Objector and the United States Armed Forces* (Bakersfield, CA: Peace Church Challenge, 2008), pp. 12f.

3 Conscience and Peace Tax International (CPTI), *Conscientious Objection to Military Service: Issues for the Country Report Task Forces United Kingdom of Great Britain and Northern Ireland*, Submission to the 91st Session of the Human Rights Committee, October 2007, p. 1 http://wri-irg.org/system/files/CPTI-HRC91-UK-en.pdf. All electronic sources cited in this chapter were accessed 31 January 2013.

4 Marjorie Cohn and Kathleen Gilberd, *Rules of Disengagement – The Politics and Honor of Military Dissent* (Sausalito: PoliPointPress, 2009), p. 36. U.S. Government Accountability Office, *Military Personnel: Number of Formally Reported Applications for Conscientious Objectors Is Small Relative to the Total Size of the Armed Forces, Report to Congressional Committees*, GAO-07-1196 (Washington, D.C.: GAO, September 2007) http://www.gao.gov/assets/270/267693.pdf

5 War Resisters' International (WRI), *Country Reports and Updates: United Kingdom*, 2 October 2008 http://www.wri-irg.org/programmes/world_survey/country_report/en/United%20Kingdom

6 Jane O'Brien, 'The Soldiers Who Can No Longer Fight', *BBC News*, 4 March 2009 http://news.bbc.co.uk/1/hi/world/americas/7900059.stm

'the numbers leaving because of Iraq [were] often obscured as they were not counted as conscientious objectors'.[7]

Applications for universal conscientious objection status are also passively or actively discouraged; the latter depends largely on the attitude of commanding officers involved in the process. In both countries there is a lack of both transparency about the existence of this right and access to information on application procedures.[8] In Britain regulations were not even available to the counselling organization *At Ease* until a few years ago.[9] The Ministry of Defence (MoD) only released the slightly different procedures of the three services in 2007 in response to a Freedom of Information Request by War Resisters International.[10]

Some US universal conscientious objection claimants encountered serious hostility in their environment. Aidan Delgado received universal conscientious objection status 18 months after applying. During that time he faced verbal and physical aggression in his unit. His commander suspected him of seeking to shirk deployment to Iraq. In Delgado's view this was due to 'the implicit judgement on them. It was as if I was saying "I'm too moral to do what you do" or that I was looking down on them because they were a bad person'.[11]

Former Specialist Augustin Augayo joined the Army in 2003 and found it difficult in training to handle or use weapons. He applied for universal conscientious objection status after deployment to Iraq as a medic. With a decision on his application pending he was redeployed, but refused to load his weapon. Instead of assigning him non-combatant tasks as the regulations recommend, he was put on guard duty. He was decorated for his service but, when his application was repeatedly rejected and he came under pressure from his command, went AWOL, then deserted, was tried, sentenced, discharged with bad conduct and never acknowledged as a conscientious objector.[12]

In Britain, the impact of commanding officers' attitudes becomes particularly obvious in the case of Ben Griffin, a former SAS soldier, who did not file for universal conscientious objector status but refused to return to Iraq on ethical and political grounds. He was released in 2005 'with a glowing testimonial from his commanding officer', describing him as a 'balanced and honest soldier who possesses the strength and character to genuinely have the courage of his

7 WRI, *Country Reports and Updates: United Kingdom*, 2 October 2008 http://www.wri-irg.org/programmes/world_survey/country_report/en/United%20Kingdom.

8 WRI, *United Kingdom of Great Britain and Northern Ireland: Human Rights and the Armed Forces*, Submission to the 91st Session of the Human Rights Committee: October 2007, September 2007, p. 1. http://www2.ohchr.org/english/bodies/hrc/docs/ngos/WarResistersInternational_UK93.pdf.

9 WRI, *Country Reports and Update*, 2 October 2008.

10 WRI, *Conscientious Objection and the British Armed Forces – Index of Papers Released under the Freedom of Information Act*, August 2007 http://wri-irg.org/news/2007/modindex-en.htm.

11 Jane O'Brien, 'The Soldiers Who Can No Longer Fight'.

12 Cohn, Gilberd, *Rules of Disengagement*, p. 32f.

convictions'.[13] He explained his refusal to go back with doubts about the legality of the war and serious ethical and legal objections to US tactics in Iraq to his commanding officer when on leave in the UK.

A military environment that is often hostile to universal conscientious objection as well as multiple conflicts of loyalty and duty, which Ficarotta and Zupan discuss in this volume, are, however, not the only problems. Universal conscientious objector status confronts dissenters who object to a specific war with an all or nothing choice. Many US resisters, who eventually discovered the options of applying for such a status, considered themselves ineligible not only because they had volunteered, but also because they did not object to war in all circumstances. Whilst the first concern is wrong, the second extends the moral dilemma for potential applicants into the future. They may consider objecting to specific operations as a duty they owe to their conscience, peers and family and be prepared to tolerate opprobrium, but since universal conscientious objector status prevents them from re-entering service they cannot follow their conscience when going to war would be compatible with their ethics.

Andre Shepherd, a US Army helicopter mechanic, had joined in 2004 in order to improve his life. He became a dissenter during deployment in Iraq. After returning to his base in Germany and further research on the justifications for and the legal status of the war, he decided not to go back. When he was scheduled to redeploy he discussed his objections with a Non-Commissioned Officer who mentioned universal conscientious objection. For Shepherd this was not an option, because 'if they're overrunning America's shores, of course [he was] going to pick up a weapon and help to defend it'.[14] He saw his choices as: 'either go to Iraq anyway or walk away from the service, because there is no other option. It's either conscientious objection … status or AWOL or just do what you're told'.[15] He sought asylum in Germany in 2008.

Shepherd's path to dissent is not unique; other cases will be discussed below. Neither are his reasons for not claiming universal conscientious objection status. On reflection he found the value system with which he had grown up irreconcilable with the tasks he was ordered to perform in Iraq, but defending the US at home was compatible with his conscience. It is difficult not to see him as a selective conscientious objector, but since he could not file for this he went AWOL.

---

13 Sean Rayment, 'I Didn't Join the British Army to Conduct American Foreign Policy', *The Daily Telegraph*, 12 March 2006 http://www.telegraph.co.uk/news/uknews/1512767/I-didnt-join-the-British-Army-to-conduct-American-foreign-policy.html.

14 Buff Whitman-Bradley et al. (eds), *About Face – Military Resisters Turn against War* (Oakland, CA: PM Press, 2011), p. 37.

15 Ibid.

Resistance Beyond 'All or Nothing'

Going AWOL, deserting or more generally resisting orders are not new phenomena. They are to some degree part of military culture and not always motivated by conflicts of conscience.[16] However, many US and British service members who became active war resisters on political or ethical grounds chose this path of separating from the military, including, in the case of US military personnel, seeking refuge in Canada.[17] Le Bouthillier in this volume examines how the Canadian courts have responded to US service personnel applying for refugee status.

The numbers of resisters or failures to follow orders to train, deploy, re-deploy or join their unit after home leave increased for years after the start of the Iraq War. By 2006 British AWOL numbers had trebled with a sharp rise in 2004 and 2005.[18] 17,470 military personnel went AWOL between 2003 and early 2010.[19] According to official US Army figures AWOL offenses increased by 4.2 per cent between 2006 and 2011, peaking in 2008; the number of offenders peaked in 2007 and decreased by 11.8 per cent until 2011; desertion offenses and deserters decreased in the same period by 24.4 per cent and 31 per cent respectively.[20] The decreases may be a result of enhanced prosecutions, which tripled between 2002 and 2006 compared with the 1997-2001 period.[21] The US Marines had also allegedly enhanced efforts to capture Vietnam era deserters in an attempt to prevent new desertions.[22]

Simultaneously, in the US, where a resistance 'infrastructure' and culture survived from the Vietnam era, recent veterans and serving personnel revived or set up associations to assist other recent veterans and war resisters still in active service and to publicly oppose especially the Iraq war. It is important to

---

16 Craig Leslie Mantle (ed.), *The Unwilling and the Reluctant* (Kingston, Ontario: Canadian Defence Academy Press, 2006).

17 Judy Keen, 'In Canada once more, U.S. troops fleeing a war', *USA Today*, 28 May 2010 http://www.usatoday.com/news/world/2010-05-23-war-resisters_N.htm.

18 Severin Carrell, 'Soldiers going AWOL have Trebled since the Invasion of Iraq', *The Independent*, 19 March 2006 http://www.independent.co.uk/news/world/middle-east/soldiers-going-awol-have-trebled-since-the-invasion-of-iraq-470486.html.

19 Michael Savage, ' More than 17,000 Episodes of Troops Going Awol since 2003', *The Independent*, 20 February 2010 http://www.independent.co.uk/news/uk/home-news/more-than-17000-episodes-of-troops-going-awol-since-2003-1905145.html.

20 Headquarters, Department of the Army, *Army 2020 – Generating Health & Discipline on the Force Ahead of Strategic Reset* (Report 2012), p. 130f. http://www.army.mil/article/72903/Army_2020__Generating_Health___Discipline_in_the_Force__Report_2012/.

21 Paul von Zielbauer, 'U.S. Army Prosecutions of Desertions Rise Sharply', *The New York Times*, 8 April 2007 http://www.nytimes.com/2007/04/08/world/americas/08iht-awol.4.5191411.html?_r=1.

22 Bill Nichols, 'Decades later, Marines hunt Vietnam-era deserters', *USA Today*, 7 March 2006 http://www.usatoday.com/news/washington/2006-03-07-deserter-side_x.htm.

note that, although they are frequently not aware of it, US service personnel have considerable freedoms to participate in public debate as long as they only represent themselves and refrain from engaging in electoral politics.[23] The modern US resistance movement is small compared to its predecessors, but the links are unsurprising. The wars in Afghanistan and Iraq also were or became politically, legally and ethically controversial. Modern resisters also share many experiences with their predecessors of the Vietnam era.[24]

Iraq Veterans against the War (IVAW) was founded in 2004 at a convention of *Veterans for Peace* (VfP), established in 1985 and active in supporting resisters and veterans.[25] IVAW expanded its scope to the Afghanistan war a few years later. It followed consciously in the footsteps of *Vietnam Veterans against the War*, as did its *Winter Soldier 2008* meeting, where resisters spoke about the conduct of operations and their objections to the wars.[26] IVAW members must be or have been serving in the military. It has chapters in all US states, including on military bases, and abroad and calls for immediate withdrawal of troops, reparation for the Iraqi and Afghan people and adequate health provisions, including trauma care, for active duty soldiers and veterans. Its membership is 'over 1,700'.[27] Compared to the size of the US military, this figure is very small and supports Ficarotta's and Zupan's point that only few make and carry through the choice of openly objecting to participating in an unjust war or a war that is carried out by unjust means.

In Britain similar resisters' groups are rare. *At Ease*, founded in 1973, is not a resistance organization, but speaks out on behalf of service personnel and provides advice on legal and disciplinary matters, family issues and for those seeking early release.[28] The British counterpart of VfP was only officially launched in April 2012.[29] Since leaving the SAS Griffin has been promoting a critical public discourse on the British military and Iraq and Afghanistan wars, although a high-court order prevents him from disclosing further details about torture of Iraqi and Afghan prisoners;[30] he also works with the virtual network *Forces Watch*.[31]

---

23 Cohn, Gilberd, *Rules of Disengagement*, pp. 77-83.

24 David Cortright, *Soldiers in Revolt – GI Resistance during the Vietnam War* (Chicago, IL: Haymarket Books, 1975).

25 Kelley Dougherty, Opening Speech – Legacy of GI Resistance, *Winter Soldier Conference*, 13-16 March 2008, National Labor College, Silver Spring, MD, 2008 http://www.ivaw.org/blog/legacy-gi-resistance/kelly-dougherty.

26 Iraq Veterans against the War, *About*, undated http://www.ivaw.org/about.

27 Brooklyn, 'Exhibition: War is Trauma Exhibition: Justseeds & IVAW', 3 December – 8 January, 2012 http://www.booklyn.org/exhibition/000577.php Current figures are difficult to establish.

28 At Ease, *About Us*, undated http://www.atease.org.uk/about.htm.

29 Veterans for Peace UK, About Us, http://veteransforpeace.org.uk/.

30 Richard Norton-Taylor, 'Court Gags Ex-SAS Man Who Made Torture Claims', *The Guardian*, 29 February 2008 http://www.guardian.co.uk/uk/2008/feb/29/military.law.

31 Forces Watch http://www.forceswatch.net/.

Former Lance Corporal Joe Glenton also became an active British resister. He went AWOL for two years in 2007, when he was due for re-deployment to Afghanistan, because he considered the war illegal, and was jailed in 2010.[32] In 2009 he had turned himself in and joined the *Stop the War* coalition. He lobbied for the withdrawal of British troops from Afghanistan, as they had 'become a tool of American foreign policy', reminding the government that it was 'obliged to ensure that the cause is just and right' and stating he 'felt ashamed' because he did not think Afghanistan had benefited from Britain's military presence.[33]

The evidence so far suggests, and more will be provided below, that US and British service personnel who have publicly questioned the legality and conduct of operations in Afghanistan and Iraq have often based their objections on moral and ethical grounds. Few, however, saw universal conscientious objection as an option. Selective conscientious objection might have provided some with a legitimate route to separation from the military, allowing them to act in accordance with their conscience now and in future. What then are the principal objections to a right to selective conscientious objection?

The State of the Debate on Selective Conscientious Objection

Some scholars, including several contributors to this volume, object to a right to selective conscientious objection because they see it as: (1) *impractical*, as it would be exceedingly difficult to establish fair and timely procedures for claimants to be dealt with appropriately;[34] (2) *unnecessary*, because the principles of moral equivalence of combatants[35] and corporate responsibility relieve individual service members of the need to justify their participation in a potentially unjust war although not of their duty to act in accordance with the laws of war on operations; (3) organizationally and politically *untenable* as the government, military authorities and commanders must be able to rely on subordinates following (legal) orders if the military is to function as the principal organization of the state capable of the professional application of force in defence of national security; in other words they cannot pick and choose which orders to follow.[36] In this volume Bergeron

32 'Awol Soldier Joe Glenton Speaks of Jail Support', *BBC News*, 26 July 2010 http://www.bbc.co.uk/news/uk-england-york-north-yorkshire-10760579.

33 Steven Morris, 'Alleged British Army Deserter Appears at Court Martial', *The Guardian*, 3 August 2009 http://www.guardian.co.uk/uk/2009/aug/03/british-army-alleged-deserter-court.

34 See Skerker and Bergeron in this volume.

35 For a debate on these and related issues see David Rodin and Henry Shue (eds), *Just and Unjust Warriors – The Moral and Legal Status of Soldiers* (Oxford: Oxford University Press, 2010).

36 For a refutation of various arguments in favour of a right or duty of all military personnel to SCO see George R Lucas, 'Advice and Dissent: 'The Uniform Perspective'', *Journal of Military Ethics*, 8/2 (2009), pp. 141-61.

and Fisher share the latter objection, arguing that a right to selective conscientious objection would undermine the constitutional construct of the social contract and the subordination of the armed forces to the civilian legitimate authority, a fundamental principle of civil-military relations in liberal democracies.

In essence, arguments against a right to selective conscientious objection seek to maintain severe limitations on the human and moral agency of service personnel. Lucas argued in rather drastic terms that this is necessary, citing fragging, the killing of officers and senior enlisted men by their subordinates in the Vietnam War, as an extreme version of dissent, to which the lower ranks can have neither a duty nor a right and which would undermine the military as a functioning organization. For him this demonstrated that '*autonomy is not the goal, it is the problem*'.[37] It is rather problematic to base an argument against granting service members a higher degree of moral agency on the comparatively small risk of such radical action. It also ignores the possibility that enabling service members to voice their dissent more constructively or release them from service in a specific war may reduce that risk.

Associated with these arguments is often the contention that it is hubris for individual service members to deem their individual conscience superior to the decisions of the legitimate authority, that is, in liberal democracies the elected government's authority over the armed forces and within the military the authority of superior officers over their subordinates. In the US, Aidan Delgado had this experience. In Britain, most of the above arguments, including the accusation of 'amazing arrogance', were deployed by the Judge Advocate in the Court Martial of Flight Lieutenant Malcolm Kendall-Smith, RAF, in 2006, although he accepted he had acted on moral grounds.[38] Kendall-Smith, a Medical Officer, had refused orders to deploy to Iraq in 2005 contesting its legality and was sentenced to eight months in prison, fined and dismissed from service; Deakin examines this and other British cases in more detail in this volume.

The case for a right to selective conscientious objection on philosophical grounds is usually linked to the demand that all service personnel have an obligation to refuse to participate in unjust wars; the conclusions to this volume examine the contributions of Imiola and Goffi on this point.[39] Others, like Skerker, concede that there may be a case in principle for selective conscientious objection, but conclude with those who object to its introduction that it is impractical, unnecessary or

37 Lucas, *Advice and Dissent*, p. 154.

38 Hyder Gulam and Mike O'Connor, 'Selective Conscientious Objection: The Court Martial of Flight Lieutenant Malcolm Kendall-Smith, RAF', *ADF Health*, 7 (Oct 2006), p. 69. http://www.defence.gov.au/health/infocentre/journals/ADFHJ_oct06/ADFHealth_7_2_68.pdf.

39 For example Jeff McMahan, *Killing in War* (Oxford: Oxford University Press, 2009); Jessica Wolfendale, 'Professional Integrity and Disobedience in the Military', *Journal of Military Ethics*, 8/2 (2009), pp. 127-40.

untenable; the moral equality of combatants must prevail. Skerker's argument rests on three core pillars:

(1) the epistemic limitations of subordinates in the armed forces which in his view do not allow military personnel, apart from the senior leadership, to make choices based on individual ethics, (2) the military as an organization takes on corporate responsibility which relieves individuals from being morally or ethically culpable as long as they execute legal orders, and (3) individuals are duty bound to execute legal orders due to the constitutional, legal and organizational relationship between the military and the state and, within the military, between individuals and the constitution and the military as an organization respectively. Essentially he suggests that the military as an organization overrides individual ethical norms on whose basis civilians have the freedom and duty to act and simultaneously absolves individual service members from applying such civilian ethical standards when executing legal orders during military service; the only basis for individual dissent is the illegality of orders.

Effectively this means service members' moral agency must be severely constrained because it is a necessary prerequisite for organizational effectiveness and because they have limited knowledge and insight. However, due to the principle of corporate responsibility they need not feel guilty for following orders they would consider ethically objectionable as civilians but which senior leaders deem operationally necessary and legal. The following sections will demonstrate through resisters' accounts, that some of the core assumptions underlying this argument are sometimes untenable. It critically assesses especially the linkage between knowledge and individual agency, how far recruits are enabled to accept the curtailment of their moral agency knowingly and by choice and the relevance of corporate responsibility for individual conscience.

Knowledge and Individual Agency

How much knowledge is enough for an individual to make an ethical choice? Perfect knowledge is never possible. Senior military and civilian officials make decisions on the basis of imperfect knowledge. As the invasion of Iraq in 2003 has demonstrated, such knowledge may be used in a manner that results in an ethically and legally contested decision. One should also not overlook, although this cannot be discussed further here, that the concepts of bounded rationality and cognitive dissonance identify limits of rationality and selectivity of information processing when actors make and uphold decisions.[40]

40 Leon Festinger, *A Theory of Cognitive Dissonance* (Stanford, CA: Stanford University Press, 1957); Barbara J. Nelson, 'Public Policy and Administration: An Overview', in Robert E. Goodin and Hans-Dieter Klingemann (eds), *A New Handbook of Political Science* (Oxford: Oxford University Press, 1996), p. 563.

Such constraints on rationality apply to any context for individual decisions, from the citizen or lowest rank in the military up to the highest levels of civilian or military public office, where structural and organizational factors add further disincentives against revising decisions even when new information would objectively make this advisable. The military context reinforces these dynamics. Its strict hierarchical organization and culture based on discipline and codes of conduct which reward efficient and effective execution of orders discourage subordinates from questioning their rationale, although they are not powerless as discussed below.

In combat it is particularly difficult for individuals to access and process information if they were to question the legality of orders. And yet, they have a duty to do so under the International Law of Armed Conflict. In cases bar the most flagrant breaches of the law, those who refuse to carry out orders they deem illegal have to defend their stance with imperfect knowledge of the finer details of the law, the operational and strategic rationale of an order, and the justifications defence and civilian lawyers will have prepared at the higher level of decision-making. Hence, the duty to resist illegal orders very likely has to be discharged on the basis of limited knowledge.

For this reason and because there is no indication how much knowledge is enough, it is not clear why individuals should not also have a right to object to technically legal but ethically questionable orders or seek to prevent the execution of orders that may have this effect. The requirement of superior knowledge for granting the right to dissent becomes even more problematic if it accords moral agency only to senior decision-makers not least because they define the Rules of Engagement.

During Winter Soldier 2008 many speakers were particularly disturbed by the gradual relaxation or break-down of the Rules of Engagement and the de-humanizing effect on soldiers of the operational environment and tactics.[41] Several veterans spoke about orders to strike buildings or fire at persons, vehicles or areas indiscriminately, which often led to the loss of civilian lives without operational gains apparent to the service members. This confronted soldiers with moral dilemmas and increased moral pressures in an already high-pressure operational context.

Jason Hurd, honourably discharged after ten years in the US Army and National Guard, chose not to act in accordance with the Rules of Engagement and thus narrowly avoided triggering a major incident. He should have prepared to fire at a car which continued to approach a check-point despite his repeated warnings. In the last minute a local man appeared and stopped the car. The driver, an elderly lady, was a greatly respected member of the local community. Hurd considered it highly likely that a major riot would have ensued had he killed her.[42]

Sergio Kochergin, formerly Marinecorps Scout Sniper Platoon, described that after casualties mounted it was left to soldiers on the ground to determine who

41 For example Hart Viges's testimony at Winter Soldier 2008 http://www.ivaw.org/blog/rules-engagement/hart-viges.

42 Jason Hurd, *Winter Soldier 2008* http://www.ivaw.org/blog/rules-engagement/jason-hurd.

was a threat. The Rules of Engagement became so relaxed that, after calling in the observation, soldiers had the authority to kill anyone they considered engaged in suspicious activity, including 'anyone carrying a heavy bag and a shovel … which was about 70 per cent of the Iraqi people … who were rebuilding their country'.[43] Kochergin was clearly haunted by the experience and ended with an apology to the Iraqi people.

Jason Lemieux was honourably discharged as a Sergeant of the US Marine corps after three tours in Iraq; he had voluntarily extended his contract by ten months. He talked about a similar relaxation of the Rules of Engagement which for his unit eventually meant 'everyone on the street was considered an enemy combatant who could be killed'.[44] Others reported that they were told to take shovels and firearms to plant on Iraqis they had killed unlawfully.[45] The latter was not identified as part of the Rules of Engagement, but the other changes meant soldiers were expected to execute technically legal orders which resulted in ethically highly problematic consequences or moral dilemmas (for example whether to follow their conscience and break the Rules of Engagement or adhere to them and act in contravention of their conscience). The question is, were they prepared for such contingencies when they decided to join the military?

Volunteers and Recruitment

Skerker concedes that a citizen who is convinced, even if he or she is wrong, that the government is launching or waging an illegal war, should not enlist and links this argument to the wider principle of not doing anything one considers morally wrong. Do potential recruits, who are civilians, understand they forfeit this right with enlistment? Are they enabled to make an informed and truly free, positive choice to join the military, rather than seeking to escape from a situation in which they perceive few options for their future, which might be considered a negative choice?

Even or especially if candidates have not reflected on their rights as civilians, perhaps because they are very young when approached by recruiters, they may take them for granted and not appreciate what the limitations mean in practice. Military life is still alien to them and their motivations for making a negative choice may override potential concerns. There is little evidence that resisters were aware of the limitations to their moral agency they would face in the military.

Some had joined because they were fascinated with technology, such as Michael Thurman who signed up for the US Air Force while still in high school,

---

43 Sergio Kochergin, *Winter Soldier 2008* http://www.ivaw.org/blog/rules-engagement/sergio-kochergin.

44 Jason Lemieux, *Winter Soldier 2008* http://www.ivaw.org/blog/rules-engagement/jason-lemieux.

45 Cohn, Gilberd, *Rules of Disengagement*, p. 51.

for patriotic reasons like William Shearer at the age of 17,[46] or because they wanted to contribute to the country's defence after the terrorist attacks of 11 September 2001.[47] When former PFC Robin Long, who came from a military family, joined after the start of the Iraq War he felt 'great', he was 'finally doing something with [his] life', 'serving his country' and doing 'something to help'.[48]

Yet, many had joined in order to turn their life around or were attracted by educational opportunities or financial and social benefits.[49] Former Staff Sergeant Camilo Mejia, now Chair of IVAW's Board of Directors, joined the Army at 19. He had attempted to continue his education whilst working and receiving federal student financial aid. When the latter was cut he signed up because he saw no other option to get a higher education.[50]

Although all were aware in principle that the benefits were counter-balanced by costs, some understood insufficiently the practical implications of joining or being sent into a warzone. They expected to deploy, but only understood what it means to kill or lose colleagues with personal experience. It would be unreasonable to expect armed forces to prepare recruits fully for such experiences, because theory will always fall short of reality. On the same grounds it would be unreasonable to expect potential recruits to anticipate how they will respond to war. Whilst many may be untroubled, some may develop ethical objections to war or specific operations.

There is ample evidence that not all US recruiters adhered to the legal and ethical constraints that exist in principle. Making false promises, such as that recruits would not be sent to war, fraudulent behaviour in order to facilitate an individual's enlistment, bullying and other misdemeanours by recruiters repeatedly attracted not only public, but also congressional attention.[51] A 2008 Government Accountability Office (GAO) report on recruiter irregularities, requested by the House of Representatives, recommended that the DOD and services enhance oversight of such irregularities.[52]

---

46 Whitman-Bradley, *About Face*, p. 127.

47 See for example the 'Rules of Engagement' Video Testimonies at Winter Soldier 2008 http://www.ivaw.org/blog/rules-engagement.

48 Whitman-Bradley, *About Face*, pp. 105 and 72.

49 See for example the cases of Benji Lewis, Ryan Johnson, Mattis Chiroux, Matt Mishler, Ghanim Khalil, Brandon Hughey in Whitman-Bradley, *About Face*.

50 Camilo Mejia, *The Road from ar Ramadi – The Private Rebellion of Staff Sergeant Camilo Mejia – An Iraq war Memoir* (Chicago, IL: Haymarket Books, 2008), pp. 15 and 24.

51 Andrew Buncombe, 'Desertion Huge Problem for US in Iraq War', *The New Zealand Herald*, 23 May 2005 http://www.nzherald.co.nz/world/news/article.cfm?c_id=2&objectid=10126890; Amy Goodman, 'Army Recruiter Suspended for Threatening High School Student with Jail Time, Sparks Bipartisan Call for Investigation', *Democracy Now!* 6 August 2008 http://www.democracynow.org/2008/8/6/army_recruiter_suspended_for_threatening_high.

52 GAO, *DOD and Services Need Better Data to Enhance Visibility over Recruiter Irregularities,* Report to Congressional Requesters, GAO-06-846 (Washington, DC: GAO, August 2006).

When public concern about questionable recruitment practices mounted, even media not associated with resistance activities or counselling addressed the problem. For example, Rod Powers, Air Force First Sergeant retired and long-time writer on the US military on About.com, a New York Times affiliate, published a multi-part article entitled 'What the Recruiter Never Told You' to shed some light on recruitment practices and the enlistment process.[53] A 2010 GAO report found that oversight of irregularities had improved, but further enhancement and greater transparency were required.[54]

Civilians entering into a contract are commonly advised to 'read the small print' and are expected to understand what it means in practice. Yet, recruits cannot understand the small print, if the constraints of military life are not explained as unambiguously as the opportunities.[55] Of course recruiters, who in the US personally benefit from high success rates, and recruitment material seek to make service life as appealing as possible. However, the British Army's application guidance notes state that service members may 'have to take a life in the act of protecting life', but not that soldiers may be killed in the course of operations or face situations in which they may have to make ethically difficult choices.[56]

The US Enlistment Document DD Form 4/1, Oct 2007, states that their status will change from civilian to military and that military law and regulations apply to them once they have joined a service, but does not explain the practical implications, mention unlimited liability or the possible need to take life.[57] Recruits are asked to confirm that they have no personal convictions, including those of a universal conscientious objector, that would prohibit them from fulfilling military duties.[58] For some who understand what this means in practice this may be true at the time of signing the contract, but change when they mature and become familiar with military life. Indeed with the provisions for universal conscientious objection the US military and legislators accept that such changes may occur.

---

53 Rod Powers, 'What the Recruiter Never Told You', *About.Com – US Military*, undated http://usmilitary.about.com/cs/joiningup/a/recruiter.htm.

54 GAO, *Military Recruiting: Clarified Reporting Requirements and Increased Transparency Could Strengthen Oversight over Recruiter Irregularities*, Report to the Subcommittee on Military Personnel, Committee on Armed Services, House of Representatives, GAO-10-254 (Washington, D.C.: GAO, January 2010).

55 See for example MoD, *Graduate Soldier – Make a Real Difference* (London: MoD, 2010) http://www.army.mod.uk/documents/general/GradSoldier_LowRes.pdf; MoD, *Are You Ready to become an Army Officer?* (London: MoD, 2008) http://www.army.mod.uk/documents/general/Officer_level_3_V_LowRes.pdf.

56 For example Her Majesty's Armed Forces, *Army Application Guidance Notes*, AFCO Form5 Army, as at 01 September 2011 http://www.army.mod.uk/documents/general/AFCO_5_Army_01_Sep_11.pdf.

57 Enlistment/Reenlistment Document Armed Forces of the United States, DD Form 4/1, October 2007 http://www.dtic.mil/whs/directives/infomgt/forms/eforms/dd0004.pdf.

58 Shubin, *The Conscientious Objector*, p. 14.

Transparency is particularly important when recruits are under age and are expected to deploy upon reaching the age of 18. In Britain minors formally have the right to leave during prescribed periods of training, but some argue they are not necessarily enabled to exercise this right.[59] In the US resisters stated that recruiters failed to inform them of their legitimate right to leave during the Delayed Enlistment Program or falsely threatened they would face legal consequences or blacklisting.[60] Resisters did not indicate awareness that 'erroneous enlistment (including contract violations, recruiter fraud, and even fraudulent enlistment)'[61] are grounds for requesting discharge.

Individual perceptions of being ill advised or misinformed during the recruitment process and at later stages of their service is relevant to the debate on selective conscientious objection. Even if recruits understand in principle the implications of the limitations on their moral agency, in weighing the risks that this will cause them conflicts of conscience they are likely to be over-optimistic if recruitment information emphasizes the thrill of service life and economic, educational or training and career prospects. Arguably they cannot make an informed decision about signing away part of the moral agency they have as civilians. Given this, it would seem that not granting them a greater degree of moral agency in service by denying them the right to selective conscientious objection reduces their human agency twice – when they join and when they serve.

This is particularly salient when one accounts for the effects of current operations on the consciences of service members. The concept of corporate responsibility may offer legal protection, but it is a weak argument for rejecting selective conscientious objection as a right. It did little to alleviate the guilt many resisters felt as a consequence of their actions or transgressions of others.

Conflicts of Conscience on Operations

That individuals began to question their reasons for joining the military or deployment to Iraq cannot be dismissed as lack of discipline or inability to cope with the hardship of life on operations. Reports from US resisters support Ben Griffin's criticism of US tactics in Iraq and his claim that US soldiers were treating

59 David Gee, *Informed Choice? – Armed Forces Recruitment Practice in the United Kingdom* (InformedChoice, 2007) http://www.informedchoice.org.uk/informedchoice/informedchoicefull.pdf ; official comments and corrections available on the website. For information on more current issues see the *Forces Watch* website at http://forceswatch.net/.

60 Rod Powers, 'What the Recruiter Never Told You Part 3 – Enlistment Process and Job Selection', *About.Com – US Military*, undated http://usmilitary.about.com/cs/joiningup/a/recruiter3_5.htm.

61 Cohn, Gilberd, *Rules of Disengagement*, p. 154.

Iraqis like 'untermenschen'.[62] They also demonstrate that this weighed heavily on the consciences of some service members.

Some had believed the Bush Administration's declared rationale for invading Iraq, but their experiences gradually led them to question the official narrative. Quite apart from a critique of the strategic rationale – the lack of the alleged linkage between 9/11 and Saddam Hussein or the inability to find Weapons of Mass Destruction – witness accounts often highlight the contradiction between the third declared reason for the invasion, to assist and liberate the Iraqi people, and the actual treatment of the Iraqi population. For example *USA Today* reported that

> Phil McDowell ... joined the Army in 2001, after the Sept. 11 ... and served in Iraq for a year. A few weeks after he was discharged in 2006, he was notified that he would be sent back to Iraq under the Army's 'stop-loss' policy ... He rejoined his unit, but he couldn't go back to Iraq. He came to Canada instead. It was a wrenching decision [which he] at first considered 'an outrageous thing to do.' But he had soured on the Iraq war: There were no weapons of mass destruction ... McDowell hated the way average Iraqis were treated by coalition forces [and] the reports of abuse of Iraqi inmates at Abu Ghraib prison by U.S. troops.[63]

Stop-loss, the involuntary extension of service by a maximum of 18 months with 12 months on operations,[64] was a particularly problematic practice, described as a 'back-door draft'.[65] Many resisters, who had unsuccessfully sought legitimate ways of separating or were due for release soon and tried to conduct themselves as morally acceptably as possible until then, went AWOL after being stop-lossed.

McDowell is also not alone with his ethical objections to the treatment of Iraqis. The empathy many resisters expressed with Iraqi people illustrates exceptionally well that their moral and ethical standard went beyond purely personal beliefs. They had been raised with these standards and norms of behaviour. In Iraq they conflicted severely with orders they received or observed others following. They were incompatible with the kind of overarching ethical belief system US regulations accept in place of belief in a supreme being as one criterion for universal conscientious objection status. One can also hardly doubt the sincerity, a second criterion for such status, of resisters who have been speaking out.

Resisters objected to their peers and superiors displaying profoundly racist attitudes towards the Iraqi population.[66] Such behaviour ranged from racist slurs

---

62 Rayment, 'I Didn't Join the British Army'.

63 Judy Keen, 'In Canada once more, U.S. troops fleeing a war', *USA Today*, 28 May 2010 http://www.usatoday.com/news/world/2010-05-23-war-resisters_N.htm.

64 Charles A. Henning, *U.S. Military Stop Loss Program: Key Questions and Answers* (Washington, DC: Congressional Research Service, R40121, 10 July 2009), p. 6f.

65 Cohn, Gilberd, *Rules of Disengagement*, p. 9.

66 See IVAW, 'Racism and War – The Dehumanisation of the Enemy', *Winter Solider 2008* http://www.ivaw.org/blog/racism-and-war-dehumanization-enemy.

or the use of otherwise disrespectful language to serious mistreatment of detainees and civilians, children and adults alike, indiscriminate killing, aimless shooting, and desecration of Iraqi dead.[67] Former Marinecorps Rifleman Vincent Emanuele reported witnessing or participating in such incidents which, he repeatedly stressed, had not been isolated. He asked the Winter Soldier audience to empathize with Iraqis who witnessed such incidents every day.[68]

Andre Shepherd became critical of US policy when he compared his experiences in Iraq with the portrayal of US military operations in the media. He felt that the Iraqi people perceived the US military presence like

> the French would look at Nazi Germany or any of the people who were overrun by the Romans … we were supposed to be the symbol of freedom and hope, so you would expect in most places you went that they would be glad to see you … I didn't want to be part of something that is the very opposite of the ideal that we were taught since we were little.[69]

PFC Robin Lang, began to doubt the official narrative in basic training and objected to the abusive language with which his instructors referred to Iraqi people. As a result he was ostracized by his peers. His doubts grew when he heard accounts from soldiers returning from Iraq and in the mainstream media. He became particularly 'angry' about stories of soldiers mutilating dead Iraqis or 'bragging' about or witnessing killings.[70] When he was unexpectedly ordered to deploy to Iraq he faced a dilemma: obey and 'participate in this atrocity and massive killing of the Arab people' or disobey, be disowned by his family, 'get a dishonorable discharge, and have a hard time even getting a job at McDonald's'.[71] He chose to escape to Canada and was subsequently disowned by most of his family.

Others empathized with the Iraqi population by arguing that Americans who found themselves under occupation, their homes invaded in the middle of the night and deprived of the only weapon they might have had for defending their home, would have responded in the same manner as the Iraqis did – fought back. The testimony of Geoffrey Millard, former Sergeant in the Army National Guard, is instructive. After a thirteen months deployment in Iraq he became engaged in anti-war activities while off active duty in 2006. His fellow unit members did not take kindly to a photograph of him at a protest event and when he joined his new unit some of its members seriously threatened him. Fearing for his life he contacted his lawyer who advised him to file for universal conscientious objector status. He 'realized [he] wasn't a CO' when he started writing and asking himself

---

67 Cohn, Gilberd, *Rules of Disengagement*, pp. 51-8; Vincent Emanuele, *Winter Soldier 2008* http://www.ivaw.org/blog/rules-engagement/vincent-emanuele.

68 Emanuele, *Winter Soldier 2008*.

69 Whitman-Bradley, *About Face*, p. 39.

70 Whitman-Bradley, *About Face,* p. 107.

71 Whitman-Bradley, *About Face*, p. 108.

> if I were an Iraqi, what would I do? If I was an American, and Chinese tanks were rolling down the streets, I probably would use that nine years of military experience and f[…] kill somebody. So I couldn't say war in all forms was bad, because when you're defending your home, you're not a bad person ... It's hard to … think about because those guys are killing my buddies, but they're still defending their homes. It's a hard place for me to be right now, I don't want them to kill my friends, but I don't want my friends to take their homeland. I wasn't a conscientious objector.[72]

Millard's conflict of conscience combines several aspects of the experiences of other war resisters: loyalty to and concern for his peers, ethical qualms about the conduct and rationale for operations in Iraq, expressed through empathy with the Iraqi people, and the conflict between seeing Iraqis as victims of, and perpetrators of threats to, US forces. He went AWOL in August 2006, but only after further reflection and engagement with accounts of Vietnam War resisters.[73] He had always objected to the Iraq War, but initially felt he could only attend protests and engage with VfP. He decided to leave when he 'stopped thinking like a soldier', demanding explanations for orders, because the 'president lied to me, my commanders lied to me, I got lied to every goddamn day in Iraq'.[74]

Perry O'Brien, Afghanistan veteran and co-organizer of Winter Soldier 2008, summed up the effects of remaining in service despite conflicts of conscience when he stated that one purpose of the event was 'the unburdening of guilt', because

> the disconnect between the [soldier's] code and what soldiers are asked to do in the war is the source of a tremendous amount of guilt that many of us carry around. Kids grow up wanting to be GI Joe and save lives. But military policy is dictating that people do terrible things … that violate their conscience, and then have the psychological burden of carrying that around, because the military says you can't talk about it.[75]

Some outspoken resisters had previously tried to effect change within the military. Many stated that they sought to avoid participating in ethically problematic actions. Some, like Camilo Mejia, increasingly and persistently tried to prevent the de-humanization of US troops and Iraqis, but were often received unfavourably.[76] The next section illustrates this further and discusses psychological effects clashes of conscience can have on service members.

---

72 Jamail, *Will to Resist*, pp. 58-9.

73 Jamail, *Will to Resist*, p. 65ff.

74 Jamail, *Will to Resist*, p. 59.

75 Cohn and Gilberd, *Rules of Disengagement*, p. 50.

76 Mejia, *Road from ar Ramadi.*

Raising Objections or Soldiering On?

Some who questioned the legitimacy of the Iraq War and objected to racist attitudes and dehumanizing behaviour of their peers and command felt unable to object at first, but eventually voiced their concerns within the military. Brandon Hughey observed such behaviour in basic training. He had

> been taught that all human life regardless of country of origin is of equal value. Just because somebody is from Iraq or the Middle East doesn't mean that their life is worth less than an American. In Basic Training you began to notice that they sort of didn't share that way of thinking, to put it lightly.[77]

At first he kept quiet to avoid becoming a target, but then raised his concerns about the legitimacy of the Iraq War with his Non Commissioned Officer. He was told to 'stop thinking so much'; when his request for discharge was declined he went AWOL to Canada. He too seems to hold sincerely an overarching set of ethical beliefs.

T.J. Buonomo, US Air Force Academy graduate and former Army Officer, discussed his critical views on the Iraq War with fellow officers and was eventually reported to his command. After an investigation he was administratively discharged, although his declared support for President Bush's impeachment could have led to a court-martial. He suspected the Army wanted to avoid a case in which he might have defended his right to free speech. He faced various conflicts of conscience. His discharge meant he could not go to Afghanistan, a war he supported, but he also felt he had done his duty to 'speak up about illegal policies and actions', when too few others were keeping quiet, and concluded that 'we need people who can really think for themselves and weigh the facts and make the best decision they can no matter how difficult a decision it is'.[78]

There is evidence of orders being disobeyed after a unit had been severely traumatized through an exceptionally large number of fatalities and casualties. The Second Platoon, Charlie Company had taken serious losses during an eleven months deployment in Baghdad in 2007. After more losses to an Improvised Explosive Device (IED) they refused to go on another patrol, as they could no longer 'function professionally' and were 'concerned that their anger could touch off a massacre of Iraqi civilians'.[79] They had been ordered to patrol an area Bravo Company had covered before its very popular Commanding Officer committed suicide in front of his unit, saying 'he had had it', and after it had lost four members to an IED. Charlie Company's Commander was not immediately made aware of the reasons for their action and the battalion labelled it mutiny. The Company

77 Whitman-Bradley, *About Face*, pp. 165-6.

78 Whitman-Bradley, *About Face*, p. 148.

79 Jamail, *Will to Resist*, p. 47.

was split up in an attempt to improve working relationships and members were punished with ineligibility for promotion or awards for two months.[80]

There are echoes of My Lai. It is not recorded that they made this link, but they acted ethically by seeking to prevent a similar outcome. Although the punishment was relatively lenient, that they were punished has direct implications for the issue at hand. Like the resisters examined above, the Company demonstrated that there are service members at all levels who reflect on their actions, think independently and are epistemologically capable of exercising moral agency, responsibly and sincerely.

Acknowledging, rather than punishing, such dissent would afford service members human dignity and reward, and thus foster, ethical behaviour. Evidence suggests that not acknowledging this can have devastating consequences for those who judge themselves by their own ethical framework, but soldier on in an operational climate that forces them to transgress these standards.

In 'None of Us were Like This Before' Joshua Phillips explores the reasons for a soldier's death presumed, but not officially acknowledged as, suicide and attributed to his inability to overcome feelings of guilt or shame about human rights abuses he had committed in Iraq.[81] Another case the US Army considers suicide (the circumstances were allegedly controversial)[82] is that of Colonel Ted Westhusing, an ethics professor at West Point who had voluntarily deployed to Iraq in 2004. In the presumed suicide note addressed to Generals Petraeus and Fil, his immediate superior, he accused the senior command of only caring about their career and failing in their professional, legal and ethical duties.[83]

These cases are anything but unique. In early 2012 suicides in the US Armed Forces reached a ten-year high and outstripped combat fatalities in the same period by more than 50 per cent; to a lesser degree the same occurred in 2008 and 2009.[84] Reasons for military suicides in the US are insufficiently understood.[85] Operational tempo, length and frequency of deployments and reluctance to seek help for mental health problems ingrained in military culture are considered important contributing factors, but health professionals have only recently begun

80 Jamail Will to Resist, pp. 48f.

81 Joshua E.S. Phillips, *None of Us Were Like This Before – American Soldiers and Torture* (London: Verso, 2010).

82 Robert Bryce, 'A Death Reconsidered', *Texas Observer*, 7 February 2008 http://www.texasobserver.org/archives/item/15282-2682-a-death-reconsidered-was-col-ted-westhusings-death-in-iraq-something-more-sinister-than-suicide Jamail cites this article, which examines whether Westhusing might have been killed because he was uncovering corrupt practices of US private contractors.

83 Jamail, *Will to Resist*, p. 166.

84 Associated Press, 'Suicides at 10-Year high in US Military', *The Guardian*, 8 June 2012 http://www.guardian.co.uk/world/2012/jun/08/suicide-rise-us-military.

85 Bryan Maxwell, 'Army Reports Highest Suicide Numbers in 2011', *Iraq and Afghanistan Veterans of America*, 19 May 2011 http://iava.org/blog/understanding-why-end-military-suicide.

to identify ethical or moral conflicts as a source of mental health problems, that is moral injury.

In 2009 Litz et al. warned that treatments for Post-Traumatic Stress Disorder (PTSD) have failed 'to pay sufficient attention to the impact of events with moral and ethical implications; events that provoke shame and guilt may not be assessed or targeted sufficiently' and sufferers were often referred to religious counsellors.[86] They stressed they did not wish to pathologize moral injury. It could only be adequately understood through 'an interdisciplinary approach (e.g. military, biological, philosophical, sociological and social psychological, legal, religious, mental health perspectives)'.[87] Long-term consequences of moral injury can be self-harming, self-handicapping and demoralization.[88] Two researchers involved in a current study at the Brite Divinity School, Fort Worth, on moral injury in veterans explicitly linked the high suicide rates to moral injury.[89]

It is not possible to estimate how many suicides might have been preventable with a right to selective conscientious objection. Even if it were a right, claimants might face as many barriers as potential universal conscientious objectors, although this would be highly undesirable as Ficarotta and Zupan argue in this volume. Yet, the available evidence does support the argument that corporate responsibility and the moral equality of combatants does not necessarily enable individuals to absolve themselves of feeling guilty for or ashamed of their transgressions or failure to prevent them. This undermines one of Skerker's core arguments against a right to selective conscientious objection.

A Right to Selective Conscientious Objection in Context

The empirical evidence demonstrates that there is a need to acknowledge that service members can have or develop moral and ethical objections to the initiation or conduct of operations which cannot be easily dismissed as unwillingness to subject themselves to military discipline. Resisters' accounts showed that many reflected critically on the rationale for the operations in which they became involved, their reasons for joining, the compatibility of the ethical framework they had acquired during their upbringing with the operational climate, and ethical questions arising from their decision to separate lawfully or illegitimately.

Recurring themes revolved around questions of honesty, trust and the dehumanization of self and other, often combined with concerns about a lack of

86 B.T Litz et al., 'Moral Injury and Moral Repair in War Veterans: A Preliminary Model and Intervention Strategy', *Clinical Psychology Review*, Vol. 29 (2009), p. 696.

87 Litz et al., 'Moral Injury', p. 696.

88 Litz et al., 'Moral Injury', p. 700.

89 Bruce Tomaso, 'Brite Divinity School wins grant to study 'moral injury' among combat veterans', *Dallas Morning News*, 17 April 2012 http://theSCOopblog.dallasnews.com/2012/04/brite-divinity-school-wins-gra.html/.

respect for humans in life and death, a culture of racism in the military, and the imposition of great moral burdens on individual service members in an extremely complex and volatile operational environment. Corporate responsibility does not absolve all service members from guilt and shame they experienced after committing, observing, or failing to prevent moral or ethical transgressions, which can lead to moral injury.

The evidence does not suggest that all service members were aware of, or prepared for, these challenges when they signed up. Fuller disclosure of the costs as well as benefits of joining would facilitate more conscious choices about enlistment. Recruits ought to be better aware of potential conflicts of conscience they might experience and the practical implications of the limitations on their moral agency. Otherwise their human and moral agency is reduced twice, when they enlist and when they serve. One might object that this would portray service life as much less attractive and make recruitment more difficult. That may indeed be the case, but would increase the likelihood that those who do join are better prepared for the complex operations US and British armed forces have faced in the past decade or so. It would, however, not eliminate the need for a right to selective conscientious objection.

For individuals, a right to selective conscientious objection would acknowledge their humanity and moral agency and enable them at least in principle to avoid severe psychological damage. It could, however, also be beneficial for the military as an organization, because it could support and promote a military and command culture that reduces the moral and ethical problems identified or equips service members to cope with them. Even without such a right the military can do a lot to improve the ethical framework of operations which many resisters identified as severely deficient. Here the onus falls on leadership at all levels.

De-humanization of self and others does not need to be accepted as an unavoidable or even necessary by-product of such operations as conducted in Iraq and Afghanistan. To the contrary,

> McMaster, when Commander of the 3rd Armored Cavalry Regiment in Iraq, forbade his soldiers from using dehumanising and derogatory language when referring to Iraqis: both because such behavior is inconsistent with the shared moral values that define a soldier's moral identity and because such behavior is potentially a verbal 'foot in the door' leading to more serious forms of abuse.[90]

The evidence has shown that this approach to leadership was not always applied. It is particularly problematic when such norms are already violated in basic training where soldiers are inculcated with core military values. When they later operate under extremely relaxed Rules of Engagement, they are less enabled to avoid or prevent abuse than those acting from a strong external ethical framework. The

90 George R. Mastroianni, 'The Person-Situation Debate: Implications for Military Leadership and Civil-Military Relations', *Journal of Military Ethics*, 10/1 (2011), p. 10.

latter are more likely to encounter ethical conflicts if they are expected to follow or condone others' compliance with technically legal but ethically questionable orders. This is problematic because of potential individual psychological consequences for the military as an institution and the effectiveness of operations, especially if they are to be considered just wars.

With reference to the theory of cognitive dissonance, Mastroianni has observed that substantial behaviour changes can result from series of incremental norm changes, as humans seek 'cognitive consistency, to see the world as orderly and comprehensible' and adjust their behaviour and attitudes accordingly.[91] Attitudinal changes can thus change behaviour and behaviour changes attitudes. The evidence has shown that this self-reinforcing cycle, sometimes supported by the Rules of Engagement, resulted in a downward spiral in units operating in Iraq. This has also been recognized by the US Army and Marine Corps, whose increased attention to research on and education in military ethics, according to Mastroianni, aim to enhance soldiers' ability to act ethically in complex combat situations. Whether the desired effect can be achieved in practice depends on active leadership because, as McMasters argued, 'it is combat stress that often leads to unprofessional or immoral behavior'.[92]

Whilst Mastroianni and McMasters focused on the importance of leadership in enabling ethical behaviour, French addressed the issue of preventing de-humanization of the self. She recounted how Medal of Honor recipient US Army Sergeant Sammy L. Davis left a class of midshipmen in no doubt that the duty of officers was not only to save their subordinates' lives, but also their humanity, because 'war is always an assault on the humanity of every individual caught up in its destructive path', an assault that 'must be resisted as much as any physical threat'.[93] If wider military culture supports and rewards the adoption and implementation of such norms by leaders at all levels, leadership and military culture reinforce each other positively. This cannot be probed in detail here, but Coleman's critical examination of the concepts of duty and loyalty and Olsthoorn's critique of integrity, all three widely considered core military values, point the way.

Coleman argued that soldiers will inevitably face legal orders that 'are inherently morally problematic' and confront them with a choice between the duty to follow orders and loyalty to their group or the military.[94] They can only act with integrity, that is adhere to the organizational ethos and comply with the law of armed conflict, and leaders need to explain the rationale for orders and the limits of loyalty. Instead of imposing a conflict of loyalties covering up colleagues' misconduct or abuse of civilians would be seen as a test of integrity. One might

---

91 Mastroianni, 'The Person-Situation Debate', p. 9.

92 McMasters cited in Mastroianni, 'The Person-Situation Debate', p. 10.

93 Shannon E. French, 'Sergeant Davis' Stern Charge: The Obligation of Officers to Preserve the Humanity of their Troops', *Journal of Military Ethics*, 8/2 (2009), p. 117.

94 Stephen Coleman, 'The Problem of Duty and Loyalty', *The Journal of Military Ethics*, 8/2 (2009), p. 112.

add that explaining the rationale for orders would also allow leaders to reflect on and question their legal *and* ethical basis; a corrective when the pressures of operations increase the risk that ethical standards slip.

Olsthoorn views the usefulness of integrity as a frame of reference for decisions more critically, although his argument does not negate the importance of explaining the rationale for orders to subordinates. Integrity, Olsthoorn argued, is a vague concept, because it can be self-referential, even self-serving, and does not necessarily lead to ethically unproblematic behaviour. It is, in his view, also a much weaker base for selective conscientious objection than moral courage, justice and respect for human life, which only Canada and Israel list as a core military value.

This analysis has shown that many resisters based their objections on values of justice and respect for human life. It argued that they showed moral courage, which was sometimes accepted as a justification for granting a service member the right to separate early. It thus suggests that they would meet Olsthoorn's criteria for granting a right to selective conscientious objection, but his three core values have wider relevance.

The analysis has also shown that service members at all levels are epistemologically capable of exercising moral agency responsibly and sincerely. Acknowledging this, it has argued, could positively reinforce ethical behaviour. It could function as a bottom-up corrective for breaches of ethical standards. Together with the leadership setting clear ethical standards of behaviour from above this would support a military culture which can reduce ethically problematic orders or behaviour.

Yet, service members should have a right to selective conscientious objection, because efforts at creating a cultural and command context which raises ethical standards cannot eliminate all risks of ethical dilemmas. This is partly due to the nature of war and character of counter insurgency operations, partly because there will be resistance within the organization, since some will not share this ethos. Individual military personnel will thus continue to develop conscientious objections to specific wars or the conduct of specific operations.

If these objections cannot be reconciled with their continuing service it is in the interest of the military, society, and individuals who are not only servants of the state but also in the care of the military, to grant a right to selective conscientious objection. The military and society should prefer avoiding the risk of moral injury to either facing the negative ramifications of its long-term consequences or the costs of healing it. Finally, a right to selective conscientious objection could play a role in organizational learning. This is an extension of Goffi's argument in this volume that, from the perspective of virtue ethics, soldiers need to develop a 'critical spirit' so they can learn from their mistakes and behave more ethically. A service member becoming a selective conscientious objector would arguably demonstrate the development of such a critical spirit at an individual level. A rise in the number of applications for selective conscientious objection would be a much more instructive indicator for the ethical 'health' of the military and politico-military decision-making processes than raw desertion and AWOL figures. The latter tend to be approached as disciplinary matters, whereas the justifications for selective conscientious objection applications would identify where and how ethical problems have arisen. They would provide the military and government with key information for corrective action within the organization and learning lessons for the conduct and initiation of operations.

Chapter 14
The Practice and Philosophy of Selective Conscientious Objection

Andrea Ellner, Paul Robinson, and David Whetham

Selective Conscientious Objection in Practice – Between Rejection and Tacit Accommodation

The evidence presented in this volume shows that national authorities deal very differently and often inconsistently with both universal and selective conscientious objection. All liberal democracies examined accept that even volunteers may develop objections to war during their service and permit applications for universal conscientious objection status, although in practice this is hard to achieve as Deakin, Ellner and Nehushtan discussed with reference to Britain, the US and Israel respectively. It is also not always a feasible course of action, as Ellner showed, because objectors do not see themselves as pacifists. Coleman and Coleman charted the evolution of Australia's recognition in law of a right to selective conscientious objection for conscripts, but not the volunteers who make up its current military. If Australia re-introduced the draft, conscripts would thus be granted greater moral autonomy than the volunteers who have chosen to serve.

Deakin, Nehushtan and Rose examined how Britain, Israel and Germany respectively deal with cases of selective conscientious objection. None of these countries recognize selective conscientious objection in law, but Britain and Germany acknowledge that conscripted or professional service members may object to participation in specific operations on grounds of conscience. They decide on a case-by-case basis whether to condone refusals to obey orders to participate in an operation individuals consider unjust, retain them in the military and permit them to serve in another post somewhere else instead, or discharge them.

Rose examines three cases in which members of the Bundeswehr applied for selective conscientious objection status. In the first the officer was released from service, the second permitted to go to an alternative posting not related to the operations to which he objected. Both officers had made their case on the basis of *ad bellum* objections. They won their cases against the same pressures service members in other countries have reported, as described in several contributions to the volume. These included pressures to obey and accusations of wilfully challenging the legitimate authority and pursuing political ends rather than the right to act in accordance with their conscience. In the third case the application, on *in bello* grounds, was denied and the Non Commissioned Officer (NCO) punished for insubordination.

Rose examines these cases against the historical background that shaped German civilian and military law and the Bundeswehr's ethos of 'Inner Leadership', whose core aim was a clear separation from the Nazi era and the Wehrmacht. Although there is just as much aversion in the Bundeswehr as in other militaries to granting publicity to conscientious objection applications, Rose concludes that the first officers' case led to a situation in which 'the primacy of politics applies only within the limits of law and statutes, beyond these the primacy of conscience rules'.

In the three Israeli court cases Nehushtan examined, conscripts were not granted greater moral agency than volunteers. Judges strongly resisted granting selective conscientious objection status, because they viewed applications as politically or ideologically motivated and feared that selective conscientious objection would spread and thus undermine national security. Only in one case did the judge recommend accommodating the applicant's conscience in his postings. Such inconsistency is not uncommon, as Rose too demonstrates with reference to German court cases. Where there is a lack of transparency, this was often seen as a desirable thing by national military and civilian authorities, as in Britain and the US, to ensure that the matter could be contained.

In the British military, Deakin found only 'sparse evidence' of strong resistance to selective conscientious objection, but very few cases went to courts martial or through the full appeals process. After high profile conscientious objection cases in World War I, the British military has sought to deny conscientious objectors publicity and the opportunity to embarrass it or the government openly. To this end it has treated selective conscientious objectors with 'an approach characterized by administrative leniency' and, if no laws had been broken, either discharged individuals 'on compassionate grounds and without acrimony' or moved them to other posts. However, it showed 'no interest in deliberating over anyone's conscience'. By treating issues of conscience as a private matter and seeking to avoid either negative consequences for its reputation and effectiveness or embarrassment to the government, the British military comes close to applying in practice Nehushtan's arguments for a right to selective conscientious objection. This is discussed in more detail below.

As in other countries, numbers of acknowledged British conscientious objectors are small, although Deakin found that protest resignations had increased over the past few years. Deakin and Ellner cite a Member of Parliament, lawyers and relevant NGOs stating that many British service members choose to go Absent Without Leave (AWOL), rather than claiming conscientious objector status, because it appears to be an easier way out of service. Matters of conscience are not the only, perhaps not even the most prevalent, reason for going AWOL, but hard figures for an estimate are not available. Hence, Ellner cautioned, there is considerable uncertainty about numbers of either conscientious or selective conscientious objectors. The same picture emerged from the US where the absolute number of approved universal conscientious objection cases is larger than in Britain, which is at least partly a function of different sizes of the armed forces. As in Britain the number of applications that were either not taken further or rejected is unknown, because no statistics are kept on such cases or they are dealt with internally.

Since some US service members who went AWOL sought refuge in Canada over the past decade, as did their forebears during the Vietnam era, Le Bouthillier's examination of Canada's legal approach to foreign conscripts and volunteers seeking refugee status on grounds of universal or selective conscientious objection is highly apposite. Canadian authorities face a particularly complex challenge. They have to decide firstly whether the conscientious objection is morally justified in the context of the orders to which applicants objected in their country of origin and secondly whether, in conjunction with an applicant's inability to seek legal recourse at home without threat of disproportionate punishment, it justifies refugee status. One particular case stood out. The Federal Court rejected as irrelevant US soldier Jeremy Hinzman's claim to refugee and conscientious objector status; he had refused to participate in the Iraq war which he considered to be illegitimate. The judge argued that as a "mere foot soldier" he would not have been excluded from protection under the International Convention on Refugees even if he had participated in an illegal war. He therefore had no *ad bellum* case to make (Deakin, Ellner and Nehushtan noted examples in which British, US and Israeli courts also denied applicants for selective conscientious objection status the opportunity to make *ad bellum* cases in court). The Canadian Federal Court demanded that Hinzman base his case only on *in bello* violations. It then rejected it, arguing that such violations occasionally occurred in war, that in the US military these were only isolated incidents and that it was thus unlikely that he would have incriminated himself with systematic *in bello* violations. Hence, there was no justification for his desertion and claim to refugee status on conscientious objection grounds.[1]

Nevertheless, having examined a succession of cases heard by the Federal Court or the Federal Court of Appeal, Le Bouthillier concludes that Canadian judges have loosened the criteria claimants, particularly deserters or draft evaders, have to meet, which may benefit selective conscientious objectors from non-democracies. He did, however, find that courts have increasingly focused on the degree of state protection available to dissenters and deserters in their country of origin and expects that claimants, including selective conscientious objectors, from established democracies will continue to encounter difficulties in making claims on the basis of insufficient state protection. Further clarification of the criteria under which Canadian courts may recognize selective conscientious objectors and grant refugee status will depend on cases reaching the Appeals Court for deliberation.

---

1 The testimony of several Iraq war resisters examined in Ellner's chapter unfortunately rather contradicts the assertion that *in bello* violations were not a routine occurrence, at least in some units.

Conscience in the Military – Private or Public?

The evidence reviewed so far suggests that in practice institutional resistance to selective conscientious objection is not as universal as might be presumed. Instead it has either been tacitly acknowledged, as in Britain, for decades or is gradually becoming a potentially legitimate reason for either re-deployment to posts not related to the operation to which individuals object, administrative discharge, or claims to refugee status. However, the evidence also suggests that there is inconsistency and disagreement about the grounds on which such claims can be made and what should result for the claimant. Amongst civilian and military authorities the fear is widespread that officially approving, or granting publicity to selective conscientious objection applications, will undermine military effectiveness, because it will foster resistance, and pose serious challenges to the legitimate authority and its ability to use force, or even to the entire system of governance.

These contentious issues are at the heart of the conceptual and philosophical debate in this volume. Before comparing the different strands of argument it is, however, worth discussing briefly Britain's attitude to, and Nehushtan's argument on, selective conscientious objection as a private matter. They stand out amongst the practical and theoretical approaches to selective conscientious objection because they occupy the middle ground of neither completely opposing it nor fully endorsing it as an official right. Both seek to accommodate selective conscientious objectors. Both seek to avoid disruption to the policy implementation process. Yet, the rationale for the British military seems to be to prevent public embarrassment and degradation of operational effectiveness, as unit members may not trust a professed selective conscientious objector to act as required in the front line, while Nehushtan argues that in principle the state has an actual obligation not to force individuals to act against their conscience lest it inflict unnecessary harm on them, regardless of operational effectiveness (at least in principle – see below).

In his analysis of the Israeli court cases Nehushtan is critical of the judges' conflation of selective conscientious objection with civil disobedience when they reject applications. He argues that individuals do not have a choice about deeply held moral beliefs and that the nature of conscience is private. It has to be distinguished from civil disobedience, which has a public dimension as it explicitly aims to change policy or laws, although he concedes that conscientious objection may have a public dimension if it is indirectly aimed at a specific policy. He further argues that the distinction between universal and selective conscientious objection is meaningless, because conscience is indivisible and there is no credible reference point to distinguish between different types, objects or scopes of objection. But individuals do have a choice whether to act on their beliefs and the state has no right to force them to act against their conscience. It has a moral obligation to accommodate them, but, he adds crucially, only as long as the number of conscientious objectors does not 'jeopardize public interest'.

Considering the strong resistance to selective conscientious objection in the Israeli courts, Nehushtan's argument in support of a right to selective conscientious objection probably goes as far as practically possible. Practical relevance is after all one of the central aims of this volume. However, treating conscientious objection as a purely private matter and conceding the right to limit the number of approvals to avoid jeopardizing public or national interest is at odds with the rationale for selective conscientious objection others, such as McMahan, have developed. They have a profoundly public aim: to prevent unjust wars. Challenging or disrupting the policy implementation process is a constituent part of this. It is most pronounced where arguments in favour of a right to selective conscientious objection are linked to the expectation that service members have a duty to object to participating in unjust wars. The salient point here is that not waging unjust wars is perceived to serve national security and the public interest. Key to this argument is, however, that the motivation for selective conscientious objection is moral, not political, even if in extremis the effect of selective conscientious objection might influence the government's ability to pursue political aims with the use of force. Several US war resisters Ellner examined explained their refusal to obey orders in these or similar terms.

There are clearly underlying tensions. Nehushtan's demarcation of selective conscientious objection accommodates both private conscience and government prerogatives, but eliminates its public rationale. Yet, granting service members a right or considering it their duty to refuse to participate in unjust wars contravenes traditional concepts of the relationship between government, military and society at several levels. The full range of contentions explored in this volume emerges from here. This concluding chapter cannot do justice to the finer points of each contribution, but we have identified themes that are common to several papers. The first section summarizes and critically engages with arguments against a right to selective conscientious objection under the heading governance and civil-military relations in liberal democracies. This is followed by two sections on individual conscience in the military context. In order to preserve the coherence of the arguments, the first of these will summarize the views of those rejecting a right and duty to selective conscientious objection before engaging with arguments in favour of a right to selective conscientious objection in the subsequent section.

Governance and Civil-Military Relations in Liberal Democracies

Bergeron, Skerker and Fisher's arguments against a right to selective conscientious objection rely largely on the concept of the social contract, which posits that citizens confer to the state certain functions, such as the monopoly of force, in the expectation that it will provide common goods, such as national security. Whether security should still be considered a common good today has been disputed,[2] but, assuming for the purpose of this discussion that it is, the debate about selective conscientious objection calls into question how and by whom that common

2 Elke Krahmann, 'Security; Collective Good or Commodity?', *European Journal of International Relations*, 14/3 (2008), pp. 379, 404.

good may be defined and protected. Constitutions of liberal democracies assign these rights to the legitimate authority, the elected government. Its privileges are in principle limited, as it is accountable to society through the latter's elected representatives, but thus safeguarded it is assumed to express the collective will and to act on behalf of community. Nehushtan's caveat that it would be legitimate for the state to curb the number of approved conscientious objectors in order to preserve its ability to protect national security is based on the same assumptions. The underlying concern is that a right to selective conscientious objection would lead to insubordination of the military to the political authority. The military would effectively be in a position to determine political decisions. Ficarotta refutes this argument, when he emphasizes that a right to selective conscientious objection applies to the individual service member, not the institution as a whole.

As Fisher highlights, models of civil military relations in liberal democracies are grounded in the basic tenet of the primacy of the political over the military. The military advises on questions about the use of force, but once the political decision has been made it is, as the servant of the state, duty bound to implement it. This reasoning is unproblematic as long as decision-makers act in the letter and spirit of the constitution and warrant the trust which the public and the armed forces are expected to place in the government, its military advisers, and the legality and legitimacy of their decisions. Several contributors have highlighted the importance of trust in civil military relations as well as within the military hierarchy. However, as Coleman and Coleman point out, this reasoning becomes problematic when that legality or legitimacy of the use of force is contested by a minority in government, a significant section of society, or the military itself, and trust is undermined; if the government overrides the views of substantial opposition, the argument that the government is protecting the common good is weakened.

This applies to questions of *jus ad bellum* and *in bello*. For Bergeron the separation of roles means that the government is responsible for *ad bellum* decisions and the military for adherence to *in bello* principles. Ultimately *ad bellum* decisions are indeed the government's prerogative and Bergeron quite rightly demands that Congress take its role of holding the government to account more seriously and actively challenge its decisions, especially if it is about to launch a war whose legitimacy and justness are widely contested. In a well-functioning democracy society, too, should also actively hold the government to account. Skerker makes a similar, equally valid point when he argues that civilians enjoy much greater moral autonomy than service personnel and thus have the right and obligation to act on their moral beliefs. The main concern here is, however, how far service personnel have a moral right or duty to question and refuse to implement government decisions to use force.

A key point is that the separation of responsibilities between the government and military is not as clear-cut as suggested and this has consequences for the degree to which the latter can be held morally accountable. At first glance Fisher seems to assume as neat a distinction as Bergeron, when he argues that *ad bellum* decisions are made at the political and strategic levels where 'respectively, the decision to go to war is taken and how it is to be won'; the conduct of the war is decided 'at the theatre and tactical level'. In contrast to Bergeron however, he accounts for the role of the senior military leadership in making *ad bellum* decision.

Fisher, Skerker and Imiola recognize that the demarcation is blurred, when they argue that the military's advisory role means senior officers who participate in this process have a duty to ensure compliance with *ad bellum* principles and raise objections if the government violates them. For Imiola they have a duty to apply due diligence, which he extends to all service personnel; his perspective is discussed in more detail below. Skerker, who rejects a general duty to refuse to participate in unjust wars and commensurate right to selective conscientious objection, concedes that it may be necessary to rethink this duty and right for senior officers who are in a position to acquire sufficient knowledge to make informed judgements whether a suggested operation complies with *ad bellum* principles. Skerker focuses on individuals' epistemological position. Fisher bases his case on the ability of senior officers to influence or control such decisions. They, and civilian decision-makers, can and should be held to account, because 'control and consent are the two key preconditions of the attribution of moral responsibility'; 'such responsibility may extend considerably further down than just the top generals'.

In contrast to Imiola and Skerker, Fisher does not support or consider a right to selective conscientious objection for these officers, as it would undermine the control of the military by the legitimate civilian authority. He recommends that they resign in order to avoid being implicated in condoning an unjust war. Resignation may seem a less controversial alternative to a right to selective conscientious objection, but it may not be less of a challenge to the legitimate authority. Elsewhere Kohn has argued that such resignations undermine civilian control profoundly,[3] Brooks that it could be perceived as grandstanding, 'the military equivalent of a vote of no confidence',[4] while Cook suggests that it may have only a marginal effect.[5] Given the latter caveat and assuming nevertheless that resignation is a viable option, if the moral reasons for it are not known publicly, the meaning of the gesture is lost and with it the corrective potential of selective conscientious objection improving the ethical behaviour of armed forces stipulated by several contributors to this volume. Imiola indeed recommends public resignation.

Whilst the military thus shares responsibility with the government in the *ad bellum* decision-making process, although it does not make the ultimate decision, governments also influence *in bellu*opractice. They co-determine with the military

---

3 Richard H. Kohn, 'The Erosion of Civilian Control of the Military in the United States Today', *Naval War College Review*, 55/3 (2002), p. 32.

4 Risa A. Brooks, 'Militaries and Political Activity in democracies', in Suzanne C. Nielsen and Don M. Snider (eds), *American Civil-Military Relations* (Baltimore: Johns Hopkins University Press, 2009), p. 220.

5 Martin L. Cook, 'Revolt of the Generals: A Case Study in Professional Ethics', *Parameters*, Spring (2008), pp. 12-13.

the Rules of Engagement or permissible practice in detaining or interrogating enemy combatants. The testimony of US war resisters examined in Ellner's contribution demonstrates how such decisions may directly expose service personnel to moral dilemmas and, if they chose to act in accordance with official legal guidance, cause them to experience shame or guilt for acting against or failing to prevent actions that contravene their deeply held moral beliefs about the just conduct of war. The same chapter also showed that a number of resisters developed conscientious objections to the justification of a war, especially the Iraq War, after they had experienced its conduct first hand or witnessed the recollections and attitudes of others who had already been on operations. This leads to the next set of issues debated in this volume, rights and obligations of service members towards their unit, the military and the state.

Individual Conscience in the Military Context – The Case against Selective Conscientious Objection

Judgements in several court cases examined in this volume and philosophical objections to a right or duty to selective conscientious objection, represented here by Fisher, Skerker and Bergeron, rest on some core assumptions about the status and obligations of professional service members. The arguments derive from the basic assumption that, when deciding to enlist, volunteers make an informed, rational choice. They can be expected to have reflected on the curtailment of their rights and the moral autonomy they enjoyed as civilians before they promised voluntarily to serve the state as represented by the legitimate political authority. They knew upon signing the contract that not only that they were they relinquishing a great deal of their moral agency, but also that they were obliged to obey all legal orders without calling them into question, if necessary including taking life and risking their life in fulfilment of their professional duties. If they objected to these limitations, they should not have enlisted.

By signing their contract they became bound by its terms. They must trust in the legitimacy of the democratically elected governments' and their military superiors' decisions, especially if that trust had been warranted in the past. Trust is the basis of an effective relationship between the military and government, the military chain of command and the cohesion and operational effectiveness of units down to the lowest level, where individuals have to be able to trust their peers to be fully committed to the mission and protect them if necessary. Other core military professional values, such as loyalty and obedience, are closely connected to it. Granting all service members a right to selective conscientious objection would allow them to pick and choose which obligations they fulfil, undermine trust vertically and horizontally in the military hierarchy, and prevent the government from exercising its legitimate right to defend the polity.

Indicating that extrapolating from past experience may not necessarily suffice to warrant trust between the government and military, Fisher and Deakin recall that the British Chief of the Defence Staff requested assurances from the Attorney General that participating in the Iraq War was legal. As such assurances were received, they argued, service members could trust that the war upon which the government was embarking did not breach *ad bellum* principles. Of course one could counter that this did not make it ethically or morally uncontroversial. For those rejecting a right to selective conscientious objection this is, however, not a matter that should concern lower ranking officers, NCOs and other ranks, because as long as they follow only legal orders both to deploy and in theatre they are protected by the moral equality of combatants or the military's corporate responsibility, even if they are deemed to be fighting an unjust war. Skerker and Fisher argue that combatants cannot be held responsible for *ad bellum* violations, because they are unable to gain sufficient knowledge to evaluate the moral legitimacy of a war. Consequently, imposing on all service members a duty to refuse to fight in unjust wars would subject them to unreasonable demands. Since the right to selective conscientious objection is tied to the duty to object, there is thus no justification or need to introduce such a right.

Finally, Fisher and Bergeron reflect on the possible consequences of granting a right to selective conscientious objection, which both view as potentially highly problematic and detrimental to the effect desired by the proponents of such a right. Fisher examines the consequences through the *in bello* lens. In essence, he suggests that removing the moral equality of combatants assumption by holding combatants responsible not only for the manner in which they fight, but also the justness of their cause, would remove important safeguards against *in bello* violations. Those who can be sure that their cause is just might take license from this knowledge and thus fight unconstrained by moral considerations in order to achieve their aim. Those whose cause is uncontrovertibly deemed unjust may do the same, because they have nothing to lose. The net effect would be a brutalization of war, which is entirely incommensurate with the aims of the Just War Tradition and, for that matter, those of the proponents of a right to selective conscientious objection.

Bergeron's concerns focus more on the effect such a right might have on the capacity of the military to recognize and safeguard against violations of the Just War Tradition. If, she argues, service members whose conscience compels them to prevent ethically and morally problematic decisions and actions were allowed to opt out of operations, only those without such a moral compass would be left. This could lead not only to an increase in the number of unjust wars, but, as in Fisher's argument, wars that would most likely be prosecuted in violation of the *in bello* considerations of the Just War Tradition. In addition, she argues, a right to selective conscientious objection would weaken the possible impact that individual objections might have. As in the case of civil disobedience, change is more likely to result if resisters or objectors have to pay a high price for taking a moral stance. Legalizing a right to selective conscientious objection would considerably decrease the penalties individuals incur, if refusals to obey remain a sanctionable disciplinary offence. A reduction of their personal sacrifice would diminish the meaning of their actions and with it their chances of affecting change.

Individual Conscience in the Military Context – The Case for Selective Conscientious Objection

Zupan, Imiola, Goffi and Ellner critically engage with the arguments against a right and duty to selective conscientious objection from a variety of perspectives. Ellner disputes the validity of the assumption that volunteers can be expected or are invariably able and enabled during the recruitment process to make a rational, informed choice to enlist and understand the sacrifices they are expected to make with regard to their moral agency once they are serving. Whilst recognizing that recruitment inevitably involves presenting service life in the best possible light, substantial evidence suggests that volunteers do not always make a free and positive choice to enlist. Clearly recruits cannot be prepared as well for life in the military before they sign their contract as they can when they are serving members and being trained for their jobs. They can only fully grasp the reality of war when on operations, and how well they cope is partly a function of the individual's disposition.

However, many join for reasons other than those imputed by authors who object to a right of selective conscientious objection. If knowledge and control are regarded as central for considering individuals to be responsible for their decisions, it is ethically at least questionable to hold responsible for the consequences of their decision to join those service members who volunteered, in extremis, under duress, on the basis of false promises, unaware of the sacrifices that accompany the benefits of service life, or because they perceived it as the only option to sustain themselves economically or improve their prospects in life. Legal avenues for seeking redress and release from service exist, but recruits are often unaware of their rights. Many arguments against a right to selective conscientious objection hinge on the construction of members of professional militaries as volunteers, but this underlying assumption is not as universally valid as suggested. Whether the state can invariably expect greater compliance of volunteers than conscripts, who should not be forced to act against their conscience, and what this means for the case in favour of a right to selective conscientious objection and the argument that service members have a duty to object to unjust wars, are questions that deserve more attention.

Like Bergeron's and Fisher's, Zupan's argument is rooted in social contract theory, but he makes a case for a duty of soldiers to disobey orders that are immoral from a practitioner's perspective. Soldiers should not assume that they serve the common good simply by adhering to their oath and being obedient to their military and political superiors without ascertaining, where possible (and he does provide a very pertinent illustration of why this might not always be possible) that the orders they have received are both legal and moral. Much of his argument links well to Goffi's and Imiola's analyses, which examine whether service members at all levels have a duty to reflect on the justness of wars or must obey orders regardless of the justness of cause. Imiola focuses on the duty and ability of service personnel to overcome invincible ignorance and what follows for the duty to obey. Goffi develops, through the lenses of Kantian deontology, utilitarianism and virtue ethics, a critique of the French *Code du soldat* and the assumption that soldiers have an absolute duty to obey.

Describing the Code as low cost deontology, Goffi finds that it discourages service members from developing their own understanding of morality and ethics in war. It merely appears to provide a set of rules of conduct and service members will be confident of acting ethically if they follow them, but it may provide a false sense of certainty derived from an assumption of equivalence between legality and morality, a point Coleman and Coleman also explore at some length. In practice, obeying legal orders can have morally problematic consequences. However, the military's insistence that they follow the set rules does not equip them to reflect on and make ethically sound decisions in complex situations on the basis of the individual's deep understanding of adequate decision-criteria.

Goffi goes even further than that when he argues that an over-emphasis on obedience, often motivated by the fear that if soldiers start to think they will disobey, reduces responsibility. A deontological approach based on rules set by the soldier's own institution is an insufficient guide for ethical behaviour in the context in which soldiers find themselves today, if they cannot gauge the consequences of their obedience. They need to understand and be able to apply utilitarian principles. This means, they 'should aim at maximizing the welfare of the greatest number, not only among [their] fellow citizens, but among all those who may be affected by [their] decisions'. However, merely training soldiers to act ethically also falls short. If they are encouraged to develop 'a critical spirit' they can 'learn from their own mistakes and develop good character'. This can ultimately result in avoiding ethical transgressions and in willingness to obey not because the institution insists on subordination, but because soldiers recognize rules or orders as ethically sound. Ficarotta and Ellner argue similarly that recruitment might even improve if volunteers were granted greater moral agency.

Goffi thus argues that, since obedience is not a moral obligation, soldiers are allowed, 'and perhaps clearly required, to refuse to fight in particular conflicts with which they do not agree', if they are to be able to protect their integrity and moral courage. Goffi and Ellner put forward overlapping arguments on the preconditions for and benefits of granting a right to selective conscientious objection. Although Ellner examined testimonies from US and British war resisters, the evidence demonstrated that objectors who went public explained their moral stance in terms not only a utilitarian, but also a virtue ethicist would recognize.

Ellner is somewhat reluctant to impose a duty to object to unjust wars on all service personnel, but her argument that a right to selective conscientious objection could support institutional learning and improve ethical behaviour along the entire chain of command complements Goffi's reasoning. One can extrapolate that nurturing a critical spirit in individuals in the manner Goffi proposes contributes to institutional learning with positive effects on the application of ethical principles in *ad bellum* and *in bello* decisions, the conduct of operations and military culture in general as Ellner envisages. Ficarotta, too, argues that military effectiveness might improve if service personnel were not uncertain about the morality of the war.

Ellner's perspective also overlaps with Imiola's focus on knowledge. Imiola argues that all service members have a duty apply moral diligence to official justifications of a war, because their oath and constitutional subordination to the legitimate political authority do not relieve them automatically of the moral responsibility not to engage in an unjust war. He disagrees with the contention that service members, except the top leaders, are inevitably in a state of invincible ignorance. They can be expected to seek knowledge that enables them to ascertain that the operations upon which they are about to embark are just. As discussed above, he argues that acquiring relevant knowledge is easier for higher ranking officers who have access to sources that informed the government's decision to use force.

Yet, in contrast to Fisher and Skerker he argues that, in liberal democracies, lower ranking officers, NCOs and other ranks too have sufficient opportunity to gain the knowledge necessary to overcome the state of invincible ignorance. In fact, with regard to the wars in Afghanistan and Iraq the problem was not that too little, but too much information was available. Imiola demonstrates that this can also pose difficulties, if the information leads to conflicting conclusions about the justness of a war. In this case, service personnel would have to be able to evaluate the credibility of the information in order to decide which arguments they consider to be most valid. That may prove impossible. Zupan's contribution illuminates these difficulties in some depth from a practitioner's perspective.

If they are left unsure about the justness of a war, they can be regarded as unable to overcome the state of invincible ignorance and have two choices. They can follow Grotius' advice and refrain from risking the violation of moral principles, which means refuse to participate, if they are not convinced that the war is just. Or, they can act on Vitoria's recommendation and 'lawfully go to war trusting the judgement of their superiors'. Most service members, he argues, are likely to do the latter, because they will be inclined to adhere to the military values of obedience, loyalty and trust. However, those who arrive at the conclusion that a war is unjust should have the right to selective conscientious objection.

Two of Imiola's arguments are particularly worth highlighting. Firstly, his conclusion that service membership does not absolve individuals from ascertaining that a war in which they are expected to participate is just and that, if it is not, 'soldiers should understand that becoming a selective conscientious objector is not an abrogation of the public trust but instead a moral obligation'. This chimes with Ficarotta's point that the public is likely to 'fight evil being done in their name', but contrasts starkly with arguments against a right to selective conscientious objection or in favour of treating conscientious objection as a private matter, access to which must not impede the government's pursuit by force of interests it has defined as a common good. In conjunction with Goffi's and Ellner's arguments, it is thus difficult to maintain that selective conscientious objection inevitably harms the national or public interest. As some of the British and US war resisters have argued, they took their stance in order to protect national security and interests.

Secondly, although Imiola attributes to all service members a duty of moral diligence, he argues that military leaders have greater responsibility in this regard.

They do have greater opportunity to acquire the knowledge required to make morally sound decisions and their subordinates trust them because of this, which is clearly of utmost importance to those who find themselves in a state of invincible ignorance after applying moral diligence to an *ad bellum* decision. The evidence examined by other contributors to this volume supports Imiola's point that most service personnel will obey orders to deploy, because they feel compelled, trust the judgement of their superiors, or are unable to ascertain whether the operations are just.

Many US and British resisters, however, developed ethical concerns on operations where they observed, felt compelled to participate in, or could not prevent, systematic *in bello* violations. Some of them chose to apply, effectively, moral diligence to the decision to the *ad bellum* case and became objectors. Others arrived at the same conclusion by comparing the officially declared objectives of the war with the immorality of the conduct and decided that the latter was incompatible with the overarching moral and ethical framework with which they had been raised and the *ad bellum* case. More junior ranks may be unable to overcome invincible ignorance before they deploy, but some can when they experience how the war is conducted. They have the intellectual capacity to arrive at their own independent assessment. Ficarotta makes the same observations.

Imiola, however, makes a further salient point about the responsibilities of senior leaders. Their subordinates also trust that, 'given the nature of their profession, [they] do their best to protect them from harm'. Ellner argues, with reference to H.R. McMaster, that this does not only mean protecting their life and limb, but also their humanity. Skerker and Fisher contend that corporate responsibility and the moral equality of combatants protect the moral status of service personnel as long as they follow legal orders. Ellner, and for that matter Goffi and Imiola, disagree. Many US war resisters testified to suffering a great deal of guilt and shame over ethical transgressions during operations. This is borne out by the identification of moral injury in increasing numbers of US service personnel. The violation of an individual's moral framework and the ensuing sense of shame can, in extremis, lead to self-harm and has been linked by some to the exorbitant suicide rate of soldiers in the US in the past year.

The US medical staff who identified moral injury linked it specifically to the types of operations conducted in Iraq and Afghanistan where, they argued, the exposure of service personnel to ethical dilemmas and morally difficult decisions is substantially greater and more problematic to deal with than one would expect in any war. This adds further weight to Goffi's argument that service personnel have to be enabled to make their own judgement about the morality of war and how to conduct it in keeping with ethical principles and Imiola's insistence that senior officers have special responsibilities to ascertain that the reasons for and conduct of military operations are morally sound. For their sake and that of their subordinates who cannot reconcile their conscience with the orders they are expected to follow a right to selective conscientious objection would seem to be a necessity.

How then might such a right be implemented? Ficarotta, for the same reasons as Imiola, assumes that only a few individuals are likely to overcome the multiple external and in-built pressures to fight. It is therefore unlikely that a right to selective conscientious objection would frequently deny the legitimate authority the ability to wage war, but it would compel the government to make a convincing case why going to war is the morally right course of action. If the number of selective conscientious objectors did impede the government's ability to use force, the *ad bellum* case would be exceptionally controversial and, as Ficarotta and our introduction have pointed out, the government and military would do well to take note. In Ficarotta's view, applying for selective conscientious objector status should be relatively informal, selective conscientious objectors should have the right to remain in the armed forces; there should be no restrictions on the timing of applications; and, if approved, their status not be recorded.

Ficarotta acknowledges that selective conscientious objectors might be labelled cowards and Deakin provided some evidence that this is not an unreasonable concern, for example when he noted that naval personnel who objected to the Falklands War were allowed to leave a ship about to deploy, not least because only 'strong' men were required on board ships going to war. However, the British military also demonstrates that a pragmatic relatively informal approach can work. It, and to some degree the German military, also demonstrate that allowing selective conscientious objectors to remain in service and re-deploying them to duties unconnected to the war to which they object is feasible. Ficarotta acknowledges that his maximalist approach to implementing a right to selective conscientious objection will likely meet military cultural and bureaucratic obstacles. Yet, he contends that concerns about negative consequences of a right to selective conscientious objection are rooted in an 'antiquated understanding of military life', whereas 'the character of the military we are stipulating for now [...] is a much different and much better one'.

Conclusion

The contributions to this volume have covered a wide range of evidence and argument both in favour and against a right to selective conscientious objection. The evidence shows that, whilst practice differs, whereas some countries do accommodate selective conscientious objectors, but treat it as a private matter, other countries reject such accommodation. Core reasons for each approach in practice are reflected in the philosophical debate. In our concluding chapter those making a case for selective conscientious objection have received slightly more space than those objecting to its introduction. This is because the former have to make a case for change and therefore have the burden of proof, which is a complex task because the case for the status quo is strong. To what degree they have succeeded remains to be seen. However, three core findings are worth highlighting.

Firstly, the case for a right to selective conscientious objection is not intrinsically linked to the argument that all service members have a duty to ensure that they only engage in just wars. However, supporters of a right to selective conscientious objection, and some who reject the case, agree that senior officers have a special duty to ensure that a case for war complies with the moral principles of the Just War Tradition. There is no consensus that this means they should have a right to selective conscientious objection. There is, however, agreement that they have greater opportunity to overcome invincible ignorance and are also in a better position to make judgements about the potential consequences of waging unjust wars for the common good of the country and society as well as the military. Their decisions must be informed by moral judgements, because many of their subordinates have to be able to trust that their senior leaders protect them from the moral harm of waging unjust war. Hence, they have a greater responsibility to act on their conscience and refuse to fight, if they cannot assure themselves that the war upon which they are about to embark is just.

Secondly, those arguing in favour of a right to selective conscientious objection are not seeking simply to impose on existing civil military relations and military culture a right which grants service personnel a degree of moral agency or even autonomy that would be disruptive to military effectiveness and civilian control of the military. Rather they see the potential and a need for profound change from the institutional down to the individual level in the way moral behaviour and ethical decisions are made and implemented. Their aim is to enhance the ability of service members to support greater compliance with the principles of the Just War Tradition in making *ad bellum* decisions and determining *in bello* behaviour. A right to selective conscientious objection can then be seen as a potential driver and an outcome of such change. It is important to note that, whilst this change can have political effects, its causes are firmly rooted in the realm of the morality and ethics of war.

The reasons for the perceived need for change then lead to the third core finding. The trigger for the renewed debate about selective conscientious objection may have been the politically contested wars in Iraq and to a lesser degree Afghanistan. However, in the course of these wars it has become increasingly apparent that they impose substantially more complex ethical and moral challenges on those tasked with their conduct. Whilst this ostensibly concerns primarily their *in bello* dimensions, it also has ramifications for *ad bellum* decisions. Although the character of future war is unlikely to replicate the experience of these two wars completely, the moral and ethical challenges are likely to be as, perhaps even more, complex.

It is for this reason that those making or influencing decisions on whether to go to war have a duty to ensure that they equip service personnel not only with adequate physical tools and military skills, but also the ability to make ethically sound decisions in extremely challenging situations and either the capacity to ascertain for themselves, or the confidence that they can trust the judgement of their superiors, that the war they are fighting is just. Some of us have argued that a right to selective conscientious objection is instrumentally linked to the military – and political – decisions-makers' ability to meet these challenges. Many others disagree. We hope that this volume has not only moved the debate on selective conscientious objection forward, but also stimulates further examination and discussion of the wider context of military ethics in theory and practice.

Bibliography

Adams, R.J.Q. and Philip P. Poirer, *The Conscription Controversy in Great Britain 1900-18* (London: Macmillan 1987).

Ambos, Kai and Jörg Arnold (eds), *Der Irak-Krieg und das Völkerrecht* (Berlin: Berliner Wissenschafts-Verlag, 2004).

Amt, Auswärtiges (ed.), *Von der KSZE zur OSZE. Grundlagen, Dokumente und Texte zum deutschen Beitrag 1993-1997* (Bonn 1998).

Aristotle, *Nicomachean Ethics*, translated by W.D. Ross (Kitchener, Ontario: Batoche Books Limited, 1999).

Aquinas, Thomas, *Summa Theologiae* (London: Blackfriars, 1964).

Arendt, Hannah, 'Personal Responsibility Under Dictatorship', in Jerome Khon (ed.), *Responsibility and Judgment* (New York: Schocken Books, 2003).

Bachmann, Hans Georg, Militärischer *Gehorsam und Gewissensfreiheit*, in Holger Zetsche and Stephan Weber (eds), *Recht und Militär. 50 Jahre Rechtspflege der Bundeswehr* (Baden-Baden 2006).

Baker, Deane-Peter, 'Making Good Better: A Proposal for Teaching Ethics at the Service Academies', *Journal of Military Ethics*, 11/3 (2012).

Bald, Detlef, *Die Bundeswehr. Eine kritische Geschichte 1955 – 2005* (München: C.H. Beck, 2005).

Baron, Marcia, 'Acting from Duty', in Allen W. Wood (ed.), *Groundwork for the Metaphysics of Morals. Immanuel Kant* (New Haven, CT, and London: Yale University Press, 2002).

Barral, Pierre, *Pouvoir civil et commandement militaire. Du roi connétable aux leaders du 20e siècle* (Paris: Les Presses de Sciences Po, 2005).

Baudissin, Wolf Graf von, 'Innere Führung und Sicherheitspolitik', in *Information für die Truppe*, 8 (1987).

Baudissin, Wolf Graf von, 'Referat auf einer Tagung für ehemalige Soldaten in der Evangelischen Akademie Hermannsburg am 3. Dezember 1951', in Klaus von Schubert (ed.), *Sicherheitspolitik der Bundesrepublik Deutschland. Dokumentation 1945 – 1977, Teil II* (Bonn: Bundeszentrale für Politische Bildung, 1978).

Baudissin, Wolf Graf von, *Soldat für den Frieden* (München: Piper, 1969).

Baudissin, Wolf Graf von and Will, Günter, 'Die schwere Geburt. Zur Entwicklungsgeschichte von Truppeninformation', in *Information für die Truppe*, 8 (1991).

Benbaji, Yitzhak, 'A Defense of the Traditional War Convention', *Ethics*, 118/3 (2008).

Benbaji, Yitzhak, 'The Responsibility of Soldiers and the Ethics of Killing in War', *The Philosophical Quarterly*, 57/229 (2007).

Bendix, Reinhard, *Max Weber: an Intellectual Portrait* (University of California Press, 1977).
Bentham, Jeremy, *An Introduction to the Principles of Morals and Legislation* (Kitchener, Ontario: Batoche Books Limited, 2000).
Brock, Peter, '"Excellence in Battle": British Conscientious Objectors as Medical Paratroopers, 1943-1946', *War and Society*, 22/1 (2004).
Brooks, Risa A., 'Militaries and Political Activity in democracies', in Suzanne C. Nielsen and Don M. Snider (eds), *American Civil-Military Relations* (Baltimore: Johns Hopkins University Press, 2009).
Burleigh, Michael, *Moral Combat, A History of World War II* (London: HarperPress, 2010).
Butcher, Tim, 'Pilot Resigns Over Lifting of Ban on Homosexuals', *The Daily Telegraph*, 28 January 2000.
Cain, Edward R., 'Conscientious Objection in France, Britain and the United States', *Comparative Politics*, 2/2 (1970).
Capizzi, J.E., 'Selective Conscientious Objection in the United States', *Journal of Church and State*, 38/2 (1996).
Carlisle, Rodney P., *Iraq War* (New York: Facts on File, Inc., 2007).
Christopher, Paul, *The Ethics of War and Peace: An Introduction to Legal and Moral Issues* (Englewood Cliffs, NJ: Prentice Hall, 1994).
Clark, Ramsey, *Wüstensturm. US-Kriegsverbrechen am Golf* (Göttingen: Lamuv Verlag, 1993).
Clifford, George, 'Legalizing Selective Conscientious Objection', *Public Reason*, 3/1 (2011).
Coady, C.A.J., 'The Status of Combatants', in Rodin, Shue (eds.), *Just and Unjust Warriors* (Oxford: Oxford University Press, 2008).
Cohn, Marjorie and Kathleen Gilberd, *Rules of Disengagement – The Politics and Honor of Military Dissent* (Sausalito: PoliPointPress, 2009).
Coleman, Stephen, *Military Ethics: An Introduction with Case Studies* (New York: Oxford University Press, 2012).
Coleman, Stephen, 'The Problem of Duty and Loyalty', *The Journal of Military Ethics*, 8/2 (2009).
Cook, Martin L., 'Revolt of the Generals: A Case Study in Professional Ethics', *Parameters*, Spring (2008).
Cook, Martin L., *The Moral Warrior. Ethics and Service in the U.S. Military* (Albany: State University of New York Press, 2004).
Cook, Martin L. and Henrik Syse, 'What Should We Mean by 'Military Ethics'?', *Journal of Military Ethics*, 9/2 (2010).
Däniker, Gustav, *Wende Golfkrieg. Vom Wesen und Gebrauch künftiger Streitkräfte* (Frankfurt am Main: Umschau, 1992).
Deakin, Stephen, 'The British Army and Homosexuality', in Hew Strachan (ed.), *The British Army, Manpower and Society into the Twenty-First Century* (London: Frank Cass, 2000).

Dennis, Peter et al. (eds), *The Oxford Companion to Australian Military History*, 2nd edn, (Sydney: Oxford University Press, 2008).
Deuteronomy 13:1
Durkheim, Emile, *Moral Education* (Mineola: Dover, 2002).
Enoch, David, 'Some Arguments Against Conscientious Objection and Civil Disobedience Refuted', *Israel Law Review*, 38/1 (2002).
Etat-major de l'armée de terre, *L'exercice du métier des armes dans l'armée de terre: fondements et principes* (Paris, Janvier 1999).
Fairweather, Jack, *A War of Chance* (London: Jonathan Cape, 2011).
Festinger, Leon, *A Theory of Cognitive Dissonance* (Stanford, CA: Stanford University Press, 1957).
Ficarotta, J. Carl, *Kantian Thinking About Military Ethics* (Farnham: Ashgate, 2010).
Ficarotta, J. Carl, 'Should Members of the Military Fight in Immoral Wars? A Case for Selective Conscientious Objection', in J. Carl Ficarotta, *Kantian Thinking About Military Ethics* (Farnham: Ashgate, 2010).
Finer, S.E., *The Man on Horseback: The Role of the Military in Politics* (1962: London: Pinter Publishers, 1988).
Fisher, David, *Morality and War – Can War be Just in the Twenty-first Century?* (Oxford and New York: Oxford University Press, 2011).
Frankfurt, H., 'Equality and Respect', *Social Research*, 64/1 (1997).
Frankfurt, H., 'Equality as a Moral Idea', *Ethics*, 98/1 (1987).
Fraser, David, *Will: Portrait of William Douglas Home* (London: Andre Deutsch, 1995).
French, Shannon E., 'Sergeant Davis' Stern Charge: The Obligation of Officers to Preserve the Humanity of their Troops', *Journal of Military Ethics*, 8/2 (2009).
Gabriel, Richard A., *The Warrior's Way: A Treatise on Military Ethics* (Kingston: Canadian Defence Academy Press, 2007).
de Gaulle, Charles, *Le Fil de l'Épée* (Paris: Perrin, 2010).
Geach, Peter, *The Virtues* (Cambridge: Cambridge University Press, 1977).
Gee, David, *Informed Choice? – Armed Forces Recruitment Practice in the United Kingdom* (InformedChoice, 2007).
'Gewissensfreiheit statt Kadavergehorsam. Freispruch für Bundeswehroffizier', in *Wissenschaft und Frieden*, 1 (2006).
Goffi, Emmanuel, *Les armées françaises face* à *la morale. Une réflexion au cœur des conflits modernes* (Paris: L'Harmattan, 2011).
Goodin, Robert E. and Hans-Dieter Klingemann (eds), *A New Handbook of Political Science* (Oxford: Oxford University Press, 1996).
Greenberg, Neil et al, 'Mental Healthcare in the United Kingdom Armed Forces', in Elspeth Cameron Ritchie (ed.), *Combat and Operational Behavioral Health* (Fort Detrick, Maryland: Borden Institute, 2011).
Grey, Jeff, *A Military History of Australia*, 3rd edn (Port Melbourne VIC: Cambridge University Press, 2008).
Groß, Jürgen, *Demokratische Streitkräfte* (Baden-Baden: Nomos, 2005).

Gross, Michael, *Moral Dilemmas of Modern War* (Cambridge: Cambridge University Press, 2010).

Grotius, 'De iure belli', in Gregory M. Reichberg, Henrik Syse, and Endre Begby, *The Ethics of War: Classic and Contemporary Readings* (Oxford: Blackwell, 2006).

Harel, Alon, 'Unconscionable Objection to Conscientious Objection: Notes on Sagi and Shapira', in 'Special Issue: Refusals to Serve –Political Dissent in the Israel Defense Forces: Comment', *Israel Law Review*, 36/3 (2002).

Hartle, Anthony E., 'Moral Principles and Moral Reasoning', in Don M. Snider and Lloyd J. Matthews (eds), *Forging the Warrior's Character: Moral Precepts from the Cadet Prayer* (Sisters, OR: Jerico, LLC, 2007).

Hayes, Denis, *Conscription Conflict* (London: Sheppard Press, 1949).

Hays, Kim, *Practicing Virtues: Moral Traditions at Quaker and Military Boarding Schools*, (Berkeley, CA: University of California Press, 1994).

Henning, Charles A., *U.S. Military Stop Loss Program: Key Questions and Answers* (Washington, DC: Congressional Research Service, R40121, 10 July 2009).

Hobbes, T., *Leviathan*, Richard E. Flatman and David Johnston (eds) (New York: Norton, 1997).

Home, William Douglas, *Sins of Commission* (Salisbury: Michael Russell, 1985).

Huntington, Samuel, *The Soldier and the State* (New York: Vintage Books, 1964).

Hurka, Thomas, 'Liability and Just Cause', *Ethics and International Affairs*, 21/2 (2007).

Hutcheson, Francis, *Inquiry into the Origins of our Ideas of Beauty and Virtue, in Two Treatises* (Indianapolis: Liberty Fund, 2004).

Irondelle, Bastien, 'L'armée française et l'éthique dans les conflits de post-guerre froide', *Critique Internationale*, 41 (2008).

Israel Law Review, 36/3 (2002).

Jackson, Robert H., *The Case Against the Nazi War Criminals*, (New York: Alfred A. Knopt, Inc., 1946).

Jones, Martin, 'The Refusal to Bear Arms as Grounds for Refugee Protection in the Canadian Jurisprudence', *International Journal of Refugee Law*, 20/1 (2008).

Jones, Martin, 'Beyond Conscientious Objection: Canadian Refugee Jurisprudence on Military Service Evasion' (CRS Working Paper Series, 2005).

Joseph, Keith, 'The Ethics of Selective Conscientious Objection', *Australian Defence Journal*, 160 (May, 2003).

Juin, Alphonse (Maréchal), *Trois siècles d'obéissance militaire (1650-1963)* (Sceaux: L'esprit du Livre, 2009).

Kagan, Shelly, 'Kantianism for Consequentialists', in Allen W. Wood (ed.), *Groundwork for the Metaphysics of Morals. Immanuel Kant* (New Haven, CT, and London: Yale University Press, 2002).

Kant, Immanuel, 'Groundwork for the Metaphysics of Morals', in Allen W. Wood (ed.), *Groundwork for the Metaphysics of Morals. Immanuel Kant* (New Haven, CT, and London: Yale University Press, 2002).

Kant, Immanuel, 'Perpetual Peace', in H. Reiss (ed.), *Kant: Political Writings*, 2nd Enlarged Edition (Cambridge: Cambridge University Press, 1991).

Kant, Immanuel, *Kritik der praktischen Vernunft* (Hamburg: 1974).
Kant, Immanuel, *Grundlegung* zur *Metaphysik der Sitten* (Hamburg: Felix Meiner, 1965).
Kantorowicz, Ernst, 'Pro Patria Mori in Medieval Political Thought', *The American Historical Review*, 56/1 (1951).
Kasher, Asa, 'Refusals: Neglected Aspects', in *Israel Law Review*, 36/3 (2002).
Kershaw, Ian, *The End –Hitler's Germany 1944-45* (London: Allen Lane, 2011).
Khon, Jerome (ed.), Responsibility *and Judgment* (New York: Schocken Books, 2003).
Kirchbach, Hans Peter von, 'Offizier im Heer der Einheit', in *Truppenpraxis*, 4 (1992).
Kohn, Richard H., 'The Erosion of Civilian Control of the Military in the United States Today', *Naval War College* Review, 55/3 (2002).
Krahmann, Elke, 'Security; Collective Good or Commodity?, *European Journal of International Relations*, 14/3 (2008).
Krulak, Charles, 'The Strategic Corporal: Leadership in the Three Block War', *Marines Gazette*, January (1999).
Kutz, Christopher, 'The Difference Uniforms Make: Collective Violence in Criminal Law and War', *Philosophy and Public Affairs*, 33/2 (2005).
Kyle, Keith, *Suez* (London: Tauris, 2011).
Lee, Nigel de, 'British Approaches to Military Obedience: Pragmatism, Operational Necessity and Moral Dilemmas', *Professional Ethics*, 10/2-4 (2002).
Leiter, B., 'Why Tolerate Religion?', *Constitutional Commentary*, 25/1 (2008).
Lewin, Ronald, Rommel *as Military Commander* (London: B.T. Batsford, 1968).
Lichtenberg, Judith, 'How to Judge Soldiers Whose Cause is Unjust', in Henry Shue and David Rodin (eds.), *Just and Unjust Warriors* (Oxford: Oxford University Press, 2008).
Linnenkamp, Hilmar and Dieter S. Lutz (eds), *Innere Führung. Zum Gedenken an Wolf Graf von Baudissin* (Baden-Baden: Nomos, 1995).
Litz, B.T. et al., 'Moral Injury and Moral Repair in War Veterans: A Preliminary Model and Intervention Strategy', *Clinical Psychology Review*, 29 (2009).
Lucas, George R., 'Advice and Dissent: The Uniform Perspective', *Journal of Military* Ethics, 8/2 (2009).
Lutz, Dieter S., 'Graf Baudissin – Reformer, Wissenschaftler, Hochschullehrer. Zum Gedenken an den Gründungsdirektor des IFSH', in Hilmar Linnenkamp and Dieter S. Lutz (eds), *Innere Führung. Zum Gedenken an Wolf Graf von Baudissin* (Baden-Baden 1995).
McMahan, Jeff, *Killing in War* (Oxford: Oxford University Press, 2011).
McMahan, Jeff, 'Debate: Justification and Liability in War', *The Journal of Political Philosophy*, 16/2 (2008).
McMahan, Jeff, 'The Morality of War and the Law of War', in Rodin and Shue (eds), *Just and Unjust Warriors* (Oxford: Oxford University Press, 2008).
McMahan, Jeff, 'Liability and Collective Identity', *Philosophia*, 34/1 (2006).
McMahan, Jeff, 'Innocence, Self-Defense, and Killing in War', *The Journal of Political Philosophy*, 2/3 (1994).

McPherson, Lionel, 'Innocence and Responsibility in War', *Canadian Journal of Philosophy*, 34/4 (2004).

Maisonneuve, Eric de la, *Le métier de soldat* (Paris: Economica, 2002).

Malament, D., 'Selective Conscientious Objection and Gilette Decision', *Philosophy and Public Affairs*, 1/4 (1972).

Mantle, Craig Leslie (ed.), *The Unwilling and the Reluctant* (Kingston, Ontario: Canadian Defence Academy Press, 2006).

Mapel, David, 'Coerced Moral Agents? Individual Responsibility for Military Service', *Journal of Political Philosophy*, 6/2 (1998).

Mapel, David, 'Innocent Attackers and Rights of Self Defense', *Ethics and International Affairs*, 18/1 (2004).

Mastroianni, George R., 'The Person-Situation Debate: Implications for Military Leadership and Civil-Military Relations', *Journal of Military Ethics*, 10/1 (2011).

Matthews, L.J. (ed.), *The Future of the Army Profession* (Boston: McGraw-Hill, 2002).

Mattox, J.M., 'The Ties that Bind: The Army Officer's Moral Obligations', in L.J. Matthews (ed.), *The Future of the Army Profession* (Boston: McGraw-Hill, 2002).

Medina, Barak, 'Political Disobedience in the IDF: The Scope of the Legal Right of Soldiers to be Excused from Taking Part in Military Activities in the Occupied Territories', *Israel Law Review*, 38/1 (2002).

Mejia, Camilo, *The Road from ar Ramadi – The Private Rebellion of Staff Sergeant Camilo Mejia – An Iraq war Memoir* (Chicago, IL: Haymarket Books, 2008).

Merkel, Reinhard, *Krieg. Was Amerika aufs Spiel setzt. Ein Präventivkrieg mag der Logik imperialer Macht entsprechen. Aber er untergräbt das Rechtsbewußtsein der Menschheit*, in Kai Ambos and Jörg Arnold (eds), *Der Irak-Krieg und das Völkerrecht* (Berlin 2004).

Messerschmidt, Manfred, *Die Wehrmachtjustiz 1933 – 1945* (Paderborn: Ferdinand Schöningh, 2005).

Milgram, Stanley, *Obedience to Authority* (New York: Perennial Classics, 2009).

Mill, John S. (1863), *Utilitarianism* (Kitchener, Ontario: Batoche Books Limited, 2001).

Miller, Seumas, *Social Action: A Teleological Account* (Cambridge: Cambridge University Press, 2001).

Moorehead, Caroline, *Troublesome People* (London: Hamish Hamilton, 1987).

Muhlnickel, Robert L., 'Just War Theory and Selective Conscientious Objection', *Review Journal of Political Philosophy*, 6/2 (2008).

Murswiek, Dietrich, *Organstreit Gauweiler/Wimmer – Antragsschrift vom 9. März* (2007).

Naumann, Klaus, *Generalinspekteursbrief 1* (Bonn 1994).

Nelson, Barbara J., 'Public Policy and Administration: An Overview', in Robert E. Goodin and Hans-Dieter Klingemann (eds), *A New Handbook of Political Science* (Oxford: Oxford University Press, 1996).

Nicholson, Hazel, 'A Disputed Identity: Women Conscientious Objectors in Second World War Britain', *Twentieth Century British History*, 18/4 (2007).
Nietzsche, Friedrich (1886), *Beyond Good and Evil*, translated by Helen Zimmern (Rockville: Serenity, 2008).
Nietzsche, Friedrich, *Human All Too Human. A Book for Free Spirits*, translated by R.J. Hollingdale (New York: Cambridge University Press, 2005).
Nolte, Georg and Heike Krieger, *Europäischer Wehrrechtssysteme* (Göttingen: Nomos, 2002).
Omand, David, *Securing the State* (London: Hurst and Company, 2010).
Orend, Brian, *The Morality of War* (Toronto: Broadview, 2006).
Orwell, George, *Homage to Catalonia* (London: Penguin Classics, 2000, first published 1938).
Øverland, Gerhard, 'Killing Soldiers', *Ethics and International Affairs*, 20/4 (2006).
Parfit, D., 'Equality and Priority', *Ratio*, 10/3 (1997).
Paz-Fuchs, A. and Sfard, M., 'The Fallacies of Objections to Selective Conscientious Objections', *Israel Law Review*, 36/3 (2002).
Pesendorfer, Michael, 'Handeln auf Befehl – kriegsvölkerrechtlich ein Rechtfertigungsgrund?', in Österreichische *Militärische Zeitschrift*, 35/5 (1997).
Phillips, Joshua E.S., *None of Us Were Like This Before – American Soldiers and Torture* (London: Verso, 2010).
Plato, *Euthyphro*.
Rae, John, *Conscience and Politics* (London: Oxford University Press,1970).
Rawls, John, *A Theory of Justice* (Cambridge, MA: The Belknap Press, 1971).
Rawls, John, 'Definition & Justification of Civil Disobedience', in Hugo Bedau (ed.), *Civil Disobedience in Focus* (New York: Routledge, 1997).
Rawls, John, *A Theory of Justice*, revised edition (Cambridge, MA: The Belknap Press, 1999).
Raz, J., *The Authority of Law* (Clarendon Press, Oxford, 1979).
Reemtsma, Jan Philipp, 'On War Crimes', in Omer Bartov, Atina Grossmann and Mary Nolan (eds), *Crimes of War: Guilt and Denial in the Twentieth Century* (New York: New Press, 2002).
Reginbogin, Herbert R. (ed.), *Die Nürnberger Prozesse: Völkerstrafrecht seit 1945* (München: De Gruyter, 2006).
Reichberg, Gregory M., Henrik Syse and Endre Begby, *The Ethics of War: Classic and Contemporary Readings* (Oxford: Blackwell, 2006).
Reilly, Major General Jeremy DSO, quoted in 'Thoughts on Command in Battle', *British Army Review*, 153 (Winter 2012).
Robinson, Paul, Nigel de Lee, and Don Carrick (eds), *Ethics Education in the Military* (Aldershot: Ashgate, 2008).
Robinson, Paul, 'Introduction: Ethics Education in the Military', in Paul Robinson, Nigel de Lee, and Don Carrick (eds), *Ethics Education in the Military* (Aldershot: Ashgate, 2008).
Robinson, Paul, 'Integrity and Selective Conscientious Objection', *Journal of Military Ethics*, 8/1 (2009).

Rodin, David, *War and Self-Defense* (Oxford: Clarendon Press, 2002).

Rodin, David, 'The Moral Inequality of Soldiers: Why *jus in bello* Asymmetry is Half Right', in David Rodin and Henry Shue (eds) *Just and Unjust Warriors* (Oxford: Oxford University Press, 2008).

Rodin, David and Henry Shue (eds), *Just and Unjust Warriors* (Oxford: Oxford University Press, 2008).

Rose, Jürgen, *Ernstfall Angriffskrieg. Frieden schaffen mit aller Gewalt?* (Hannover; Ossietzky, 2009).

Rose, Jürgen, 'Die Sanitäterin im Krieg. Der Fall der Bundeswehrsoldatin Christiane Ernst-Zettl und der Bruch der Genfer Konventionen am Hindukusch', in *ZivilCourage*, 1, February/March (2008).

Rose, Jürgen, 'Aufklären, damit die anderen bomben können. Dokumentation. Antrag des Oberstleutnants Jürgen Rose, von allen dienstlichen Aufgaben bei einem Tornado-Einsatz in Afghanistan entbunden zu werden', in *Freitag – Die Ost-West-Wochenzeitung*, 12, 23 March (2007).

Rose, Jürgen, *Dienstliche Erklärung vom 1. Mai 2006*, documented in Jürgen Rose, *Soldaten-Verweigerung gegen völkerrechts- oder grundgesetzwidrige Befehle*, FriedensForum – Zeitschrift der Friedensbewegung, 3 (2006).

Rose, Jürgen, *Gewissensfreiheit statt Kadavergehorsam. Freispruch für Bundeswehroffizier*, in *Wissenschaft und Frieden*, 1 (2006).

Rose, Jürgen, *Ignoriertes Urteil? Wie die Bundeswehr mit der Leipziger Entscheidung zur Gewissensfreiheit umgeht*, script for NDR Info Das Forum, 'Streitkräfte und Strategien' by Andreas Flocken, 14 January (2006).

Rose, Jürgen, *Gehorsam bis zum Hindukusch? Das Grundgesetz verbietet die Vorbereitung von Angriffskriegen. Trotzdem wird ein Bundeswehrsoldat, der sich weigert mitzumachen, degradiert*, in *Sicherheit und Frieden – Security and Peace*, 22/3 (2004).

Rose, Jürgen, *Gewissensnöte – Bundeswehr-Soldaten, die sich Auslandseinsätzen verweigern*. script for NDR Info Das Forum, 'Streitkräfte und Strategien' by Andreas Flocken, July 10 (2004).

Rose, Jürgen, 'Vom Wesen und Gebrauch zukünftiger Streitkräfte', in *Truppenpraxis*, 4 (1993).

Rousseau, J., *The Social Contract & Discourses* (London: Everyman, 1973).

Sagi, Avi and Ron Shapira, 'Civil Disobedience and Conscientious Objection', *Israel Law Review*, 36/3 (2002).

Sandburg, Carl, *The Complete Works of Carl Sandburg*, revised and expanded edition (New York: Harcourt Brace Jovanovich, 1969).

Schirmer, Gregor (ed.), *Die Nürnberger Prinzipien – ein Umbruch im Völkerrecht*, Bulletin für Faschismus- und Weltkriegsforschung (Academic Half-Year Journal, ed. in cooperation with the Berliner Gesellschaft für Faschismus- und Weltkriegsforschung e. V.) (Berlin: Edition Organon, 2006).

Schubert, Klaus von (ed.), *Sicherheitspolitik der Bundesrepublik Deutschland. Dokumentation 1945–1977, Teil II* (Bonn: Bundeszentrale für Politische Bildung, 1978).

Shubin, Daniel H., *The Conscientious Objector and the United States Armed Forces* (Bakersfield, CA: Peace Church Challenge, 2008).

Shue, Henry, 'Do We Need a Morality of War?', in Rodin and Shue (eds), *Just and Unjust Warriors* (Oxford: Oxford University Press, 2008).

Smith, Hugh, 'Conscientious Objection to Particular Wars: Australia's Experience during the Vietnam War 1965 – 1972', *War & Society*, 8 (1990).

Smith, Hugh, 'Conscience, Law and the State: Australia's approach to Conscientious Objection since 1901', *Australian Journal of Politics and History*, 35 (1989).

Smith, Rupert, *The Utility of Force* (London, Allen Lane, 2005).

Snider, Don M. and Lloyd J. Matthews (eds), *Forging the Warrior's Character: Moral Precepts from the Cadet Prayer* (Sisters, OR: Jerico, LLC, 2007).

Sola, Andrew, 'The Enlightened Grunt? Invincible Ignorance in the Just War Tradition', *Journal of Military Ethics*, 8/1 (2009).

Steinhoff, Uwe, 'Debate: Jeff McMahan on the Moral Inequality of Combatants', *The Journal of Political Philosophy*, 16/2 (2008).

Sternberg, Mark R. von, *The Grounds of Protection in the Context of International Human Rights and Humanitarian Law: Canadian and United States Case Law Compared* (The Hague; New York: Martinus Nijhoff, 2002).

Strachan, Hew (ed.), *The British Army, Manpower and Society into the Twenty-First Century* (London: Frank Cass, 2000).

Strenski, Ivan, *Contesting Sacrifice: Religion, Nationalism, and Social Thought in France* (Chicago, IL: The University of Chicago Press, 2002).

Takemura, H., *The International Human Right to Conscientious Objection to Military Service and Individual Duties to Disobey Manifestly Illegal Orders* (Heidelberg: Springer, 2009).

Tarvis, Carol and Elliot Aronson, *Mistakes Were Made (But Not by Me): Why We Justify Foolish Beliefs, Bad Decisions, and Hurtful Acts* (San Diego: Harcourt, 2007).

Thomas, Ward, 'Unjust War and the Catholic Soldier', *Journal of Religious Ethics*, 35/3 (2007).

Time Magazine, 'Cyprus: Grounded Bomber', *Time Magazine*, 11 February 1957.

Tootal, Stuart, *Danger Close* (London: John Murray, 2009).

Vitoria, Francisco de, 'On the Law of War', in Gregory M. Reichberg, Henrik Syse, and Endre Begby, *The Ethics of War: Classic and Contemporary Readings* (Oxford: Blackwell, 2006).

Vitoria, Francisco de, 'On the Law of War', in Anthony Pagden and Jeremy Lawrance (eds.), *Vitoria: Political Writings* (Cambridge: Cambridge University Press, 1991).

Wall, Peter, 'The Ethical and Legal Challenges of Operational Command', in David Whetham (ed.) *Ethics, Law and Military Operations* (Basingstoke: Palgrave Macmillan, 2010).

Walters, Doug, *The Doug Walters Story* (as told to Ken Laws), (Adelaide: Rigby, 1981).

Walzer, Michael, 'Response to McMahan's Paper', *Philosophia*, 341 (2006).

Walzer, Michael, *Just and Unjust Wars: A Moral Argument with Historical Illustrations* (New York: Basic, 1977).
Walzer, Michael, *Obligations* (Cambridge, MA: Harvard University Press, 1970).
Wells, Donald A. (ed.), *An Encyclopedia of War and Ethics* (Westport, CT: Greenwood, 1996).
Wertheimer, R., 'Reconnoitering Combatant Moral Equality', *Journal of Military Ethics*, 6/1 (2007).
Whetham, David (ed.) *Ethics, Law and Military Operations* (Basingstoke: Palgrave Macmillan, 2010).
Whetham, David, 'The Just War Tradition: A Pragmatic Compromise' in David Whetham (ed.) *Ethics, Law and Military Operations* (Basingstoke: Palgrave Macmillan, 2010).
Whitman-Bradley, Buff et al (eds), *About Face – Military Resisters Turn against War* (Oakland, CA: PM Press, 2011).
Wing, Ian, 'Selective Conscientious Objection and the Australian Defence Force', *Australian Defence Force Journal*, 137 (1999).
Wissenschaft und Frieden, 1 (2006).
Wolfendale, J., 'Professional Integrity and Disobedience in the Military', *Journal of Military Ethics*, 8/2 (2009).
Wood, Allen W. (ed.), *Groundwork for the Metaphysics of Morals. Immanuel Kant* (New Haven, CT, and London: Yale University Press, 2002).
Wrage, Stephen, 'Captain Lawrence Rockwood in Haiti', *Journal of Military Ethics*, 1/1 (2002).
Yingling, Paul, 'How Congress Has Abdicated Its War Powers', *Armed Forces Journal*, February (2010).
Zetsche, Holger and Stephan Weber (eds), *Recht und Militär. 50 Jahre Rechtspflege der Bundeswehr* (Baden-Baden: Nomos, 2006).
Zielbauer, Paul von, 'Army is Cracking Down on Deserters', *The New York Times*, 9 April 2007.
Zohar, Noam, 'Collective War and Individualistic Ethics', *Political Theory*, 21/4 (1993).
'Zonshein v. Judge Advocate General', *Israel Law Review*, 36 (2002).
Zupan, Dan, 'A Presumption of the Moral Equality of Combatants', in Rodin, Shue (eds), *Just and Unjust Warriors* (Oxford: Oxford University Press, 2008).

Index